THE
ALCHEMIST'S
KITCHEN

EXTRAORDINARY POTIONS
& CURIOUS NOTIONS

Guy Ogilvy

Walker & Company
New York

WOODEN
BOOKS

*This book is dedicated to Sidi Ibrahim Izz al-Din and
Acharya Manfred Junius, with sincere thanks for all that they shared.*

For those seeking to explore Alchemy further
Alchemy: Science of the Cosmos Science of the Soul by Titus Burckhardt and
The Practical Handbook of Plant Alchemy by Manfred M. Junius are indispensible guides.

Alexander von Bernus' *Alchemie und Heilkunst,*
The Weiser Concise Guide to Alchemy by Brian Cotnoir and
Mircea Eliade's *The Forge and the Crucible* are also highly recommended, as are
The Golden Game by Stanislas Klossowski De Rola (for Alchemical emblems) and
Adam McLean's Alchemy website at *levity.com/alchemy* (for pretty much everything,
including most of the images in this book).

For those inspired by the practical appendices *Caveman Chemistry* by Kevin M. Dunn
and *Formulas for Painters* by Robert Massey are both excellent books.

Many thanks to Sir Francis Melville, for access to his extraordinary library,
to Daud Sutton and John Martineau for their editorial assistance,
and to Victoria, for keeping the kitchen cooking.

*Caveat: Alchemy can be extremely dangerous. Explosions and poisonings are commonplace.
Some of the processes described in this book may be unlawful in some jurisdictions;
they are performed at your own risk.*

TABULA SMARAGDINA
The Emerald Tablet of
HERMES TRISMEGISTUS

This is the truth, the whole and certain truth, without a word of a lie.

That which is above is as that which is below,
And that which is below is as that which is above.
Thus are accomplished the miracles of the One.

And as all things come from the One, through the mediation of the One,
So all things are created by this One Thing through adaptation.

Its father is the Sun; its mother the Moon.
The Wind bears it in its belly; the Earth nurtures it.

It engenders all the wonders of the Universe.

Its power is complete when it is turned to Earth.

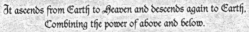

Separate the Earth from Fire, the subtle from the gross,
Gently and with great ingenuity.

It ascends from Earth to Heaven and descends again to Earth,
Combining the power of above and below.

Thus you will achieve the glory of the Universe
and all obscurity will flee from you.

This is the power of all powers, for it overcomes every subtle thing
And penetrates every solid thing.

Thus was all the world created.

And thus are marvellous works to come, for this is the process.

Therefore am I called Thrice-Greatest Hermes,
for I am master of the three principles of universal wisdom.

This concludes what I have to say about the work of the Sun.

CONTENTS

Published by
Walker Publishing Company, Inc., New York
Distributed to the trade by
Holtzbrinck Publishers

Printed on recycled paper.

Library of Congress Cataloging-in-Publication Data
has been applied for.

ISBN-10: 0-8027-1540-0
ISBN-13: 978-0-8027-1540-1

Visit Walker & Company's Web site
at www.walkerbooks.com

First U.S. edition 2006

1 3 5 7 9 10 8 6 4 2

Designed and typeset by
Wooden Books Ltd, Glastonbury, UK

Printed in the United States of America

INTRODUCTION

The Royal Art of alchemy remains one of the most enduring and baffling human enterprises. It is called the Royal Art, because it was practiced by, or on behalf of, kings and princes as far back as the legendary "Yellow Emperor" Huang-Di [ca. 27th C. BC] and as late as the 17th century, when the Holy Roman Emperor Rudolf II devoted much of his time to its study.

But what exactly is alchemy? Even the origin and definition of the word are obscure. In China it represents the quest for immortality, in India it is the art of making medicines, while in the West it is associated with the quest for the *Philosopher's Stone*, which transmutes base metals into gold. Alchemy is all these things, and more besides.

Alchemists are perfectionists, seeking to perfect everything they work with, including, and most especially, their own souls. Quite how they go about their business – and why – is the subject of this book. Confining ourselves in the main to the Western alchemical tradition, we will look at the philosophy and principles that guide alchemists, the materials they work with and the obscure, but fascinating, language and symbols they use to express their art. But be prepared – alchemy is not easy to make sense of. It is full of pitfalls and paradoxes. Getting to grips with it requires imagination and concentration.

THE SECRET ART
in pursuit of gold

To open a typical alchemical work is to be confronted by a bewildering mixture of baffling text and extraordinary, incongruous imagery. Directions for making *The Stone* are couched in arcane terminology and accompanied by pictures of strange hermetic symbols or magical dreamscapes in which royal families act out a bizarre soap opera involving marriage, discord, infanticide, regicide, hermaphrodism and graveyard sex, accompanied by a fantastic bestiary of dragons, green lions, unicorns, phoenixes, and salamanders. The writers of such books often have strange Latinized pseudonyms and lives full of intrigue and mystery.

The 17th century Polish alchemist Michael Sendivogius is a good example. Twice escaping torture and imprisonment from rapacious German princes, he performed transmutations for Rudolf II, whom he served for many years as physician and advisor, and he was also possibly the first person to isolate oxygen.

Where, amid all this confusion, is the poor neophyte to start? If you want to be an alchemist, or at least get stuck into a few alchemical potions, you have first to think like an alchemist. Remarkably, alchemists of all ages and all lands tend to share the same vision. They may have a strange way of saying things, but at least they are saying the same thing. More or less. They all believe that we, and almost everything else, are not all that we could be. Except gold. The story of alchemy, at least in the West, is, in a nutshell, the story of gold and our relationship with it. This story begins, fittingly, in a mythic *Golden Age* at the dawn of our time.

Two fishes representing the beginning and the end of the Work, also symbolizing the twelfth and final zodiacal sign of Pisces.

The dragon represents both the passional lower self, and the unrefined prima materia that must be subdued in order for the Work to begin.

Two fighting eagles represent the antagonism of Spirit and Soul, the alchemists' Sulphur and Mercury, prior to their reconciliation.

Once refined, the warring Principles are ennobled and transformed, and ready to approach each other in dignity and harmony.

Ecce Homo
in the beginning

Consciousness dawns on Primordial Man. He finds himself bathed in the light of the fiery Sun; standing upon the earth; breathing air. Ecce Homo. His wonderment is encroached upon by an ever increasing need. He thirsts for water. Mercifully it calls and attracts him. Fire, earth, air and water. Man in his element. Night falls on the Golden Age. Engulfed by darkness and a sense of loss, Primordial Man is confronted by duality: night and day; light and darkness; heat and cold. In the Sun's absence he yearns for its light, its fire. But until he is able to steal it, fire remains the property of the gods. It falls to earth as thunderbolts and blazing lumps of meteoric iron, while from below it erupts balefully from volcanoes and, sparked by sunlight through crystal, rages destructively through forests.

Meanwhile, at the water's edge, he learns of depth and reflection and finds all that he needs – to slake his thirst, fill his belly and fire his imagination. He is drawn by richly colored clays – red and yellow ochres, the color of blood, fire and sun; white kaolin the color of bones, teeth and moon; black clay the color of night. He sees them, he touches them. They color his fingers, he paints his body. Armed with the colors of the *Great Work*, the colors of the races of man, he can create likenesses of things. By calling a thing to mind, its spirit manifests itself through his hand and palette. In this way he gains power over his rivals and allies: the spirits he feels around him, his fellow man, and the animals he hunts for hide, flesh, and bone.

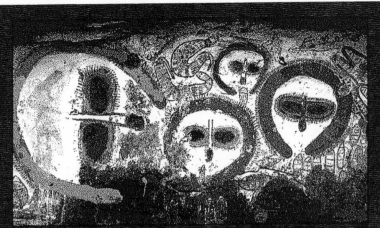

Wandjina paintings from NW Australia ~ eerie portraits of subtle beings who share our world. Legend tells that the red ochre mined for such paintings formed when the blood of ancestor spirits fell to the ground. Red ochre, white kaolin and black charcoal are still used to this day.

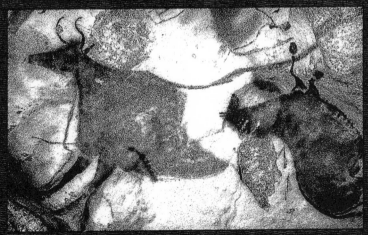

Palaeolithic cave painting from Lascaux, France ~ simple mineral pigments such as finely ground black manganese ore and haematite (the same iron oxide found in red ochre), were sprayed from the mouth and dabbed on with fingers to embody the spirits of animals.

FIRE AND METALS
from golden age to iron age

As well as garish clays, the streambeds also bore lumps of native gold – bright and shiny, the color of the sun. Precious. Intriguingly heavy, hard, but not too hard, it could be worked with stone and fashioned into the finest artefacts, things that never crumbled or decayed.

Gold, however, was not the only metal immediately available. Meteoric iron was also found lying naked on the ground. Dull, hard and unworkable, it nevertheless had a gold-like ring to it, and despite its earthy appearance was believed to have fallen from the sky. This conferred on it an awesome, mysterious, celestial quality. Artefacts fashioned from meteoric iron had magical qualities, but before this metal could be effectively worked mankind had first to become master of fire. Similar substances were encountered hidden, half-formed in their matrix of rock, as the goldstreams were pursued into the earth. Fire would yield them up.

Fire transforms things. It transformed our lives. It allowed us to bake the river clays into vessels for cooking, carrying and storing; into bricks for building furnaces that could create sufficient heat to extract metal from rock and mould it into all manner of tools, first hammers and tongs, then blades – ploughshares and weapons.

Once fire had been sufficiently mastered, the struggle for dominion over the earth was on. Despite being the symbol of perfection and permanence, gold had set us on the path of change, the metal road leading to the industrial machine age, nuclear technology and the Philosopher's Stone.

BACK TO NATURE
the principles of life

For all his technological prowess man remains at the mercy of the elements, as much a part of the living environment as everything else on Earth. Alchemists accordingly believe that *Nature* is the principle that unifies all things and governs their individual natures, and they also recognize that everything in Nature is reflected in ourselves.

To the alchemist, the universal life-giving principle within Nature is *spirit*, while the unique essence of each thing is its *soul*. These, together with the third principle, the *body*, form the *tria prima*. The easiest way to approach this central theme is to turn to the willing guidance of the plant kingdom, whose three principles are easily identified.

Plant alcohol, or ethanol, is called *spirit* because that's what it is – the spirit of a plant; the same whether made from grapes, grain or mandrake roots, thus the *universal* principle of the vegetable kingdom. The *individual* essence of a plant – its *soul* – is found in its essential oil (a rose has many names, but its fragrance is unique). The *body*, thirdly, is an invisible *salt*, extracted from the plant's ashes by separating the "subtle from the gross," as we shall see later.

The salt of plants obligingly acts as a bridge between the vegetable and mineral kingdoms, the entry point to mineral alchemy, the operations of which mysteriously reflect the processes within the transforming soul of the alchemist. The key to these processes is the interaction of the tria prima, so let's take a closer look at them and the rich symbolizm with which alchemists clothe them.

"The Wind bears it in its belly." Personified here, the wind carries the mysterious "it" of the Emerald Tablet, which all alchemists seek to identify.

"Follow Nature" advised Paracelsus. Here we see the alchemist with lantern, stick, and spectacles, attempting to follow in her footsteps.

The salamander is the elemental spirit of Fire, which must be understood in all its subtlety so as not to confuse it with "vulgar fire."

Two alchemists observe the celestial movements and relationships in order to determine the most propitious moment to begin the Work.

SULPHUR AND MERCURY
the reconciliation of opposites

In alchemical terminology soul and spirit go by the names of *Sulphur* and *Mercury*. Quite distinct from common sulphur and quicksilver, they are instead the first principles of being, originating at the dawn of creation. Together they form a duad, a polarity of complementary, but opposing forces that must be reconciled. Like the yin-yang symbol, they not only reflect each other, but contain the starting point of the other. Hence the myriad paradoxes that make classical alchemical recipes so confusing.

Sulphur, as the soul, is consciousness, the individual spirit. The hot, dry, fiery, masculine principle, it is the active, engendering seed, called Sol and the Father of the Stone. It is form – *eidos* – the idea of a thing, as opposed to matter, the expression of the idea. Its symbols include the Sun, the stag and the Red Lion. In its unpurified state it is the red man, who quarrels with the white woman. When exalted, or perfected, he becomes the Red King.

Mercury, as the spirit, is the life-force, the Universal Soul in all things. It is passive, feminine, cold and watery, the eternal feminine, the *Prima Materia* – first matter, the matrix, the mother of all things. Unrefined, Mercury is symbolized by the dragon, the serpent, the Green Lion, and the white woman who, when exalted, becomes the White Queen or the White Lion, the unicorn, or the Moon called Luna and Diana, the virgin divinity in Nature.

The third principle, *Salt*, acts as mediator between Sulphur and Mercury. It is the spark between them, the child of the union, the harmonizing balancing point of their polarity.

Johann Daniel Mylius, *Philosophia Reformata*, 1622

Michael Maier, *Atalanta Fugiens*, 1618

THE CHYMICAL WEDDING
the marriage of the sun and moon

Mankind is a paradoxical creature, full of contradictions and warring passions. The spirit wants to rule the world, the soul just wants to be happy. Their conflict is often symbolized in alchemy by a man with drawn sword and a woman with an eagle or by two fighting animals, such as eagles, or the dog and bitch, whose fighting leads to frenzied copulation and death, symbolizing the fatal futility of the love-hate relationship.

To escape this brutal cycle, and before harmony can be achieved, the inessential must be removed – the subtle separated from the gross, for as long as spirit and soul are chained by the material state they cannot be freed. The lesson is simple – if we identify too strongly with our physical selves we are doomed to share the body's death, so this false identity must be sacrificed, destroyed to reveal the true self. Likewise, a seed cannot flourish until the outer husk has rotted and fallen away. The substance that breaks down the material body is *Philosophic Mercury*, a rarified spiritual solvent, the preparation of which presents the laboratory alchemist with one his greatest challenges.

Released from their limited state, both Principles can be purified and reconciled, whereafter their sacred union can occur. This is the chemical wedding of the Red King and the White Queen. The child of their union is the transcendent androgenous child, spirit ensouled; the immortal spiritualized soul described in the title of an anonymous 18th century alchemical tome as *The Hermaphrodite Child of the Sun and the Moon*.

Aggressive unrefined Sulphur, here personified as the Red Man, attempts to indulge his unbridled passion by forcing himself on a reluctant Mercury, the unyielding White Woman. He needs to improve his technique.

The battle of the sexes is expressed here as a cannabilistic frenzy of murderous copulation, resulting in the death of both from which the alchemist must ressurect them.

Happily united, the lovers come together in the nuptial bedchamber as the Sun and Moon look happily on and the lovers' spirits soar in harmony. The alchemist awaits the result of their union.

Sulphur and Mercury as Sol and Luna find a watery cave where they can embrace in secret. Luna conceives the Hermaphroditic child who will later emerge, fully formed, from the Mercurial water.

THRICE GREAT HERMES
tricksy psychopomp

The genius of Western alchemy, and mercurial guide to all alchemists, is the legendary Hermes Trismegistus. Although considered one of the ancients and equated by the Arabs with the prophet Idris (Enoch), he combines the divine qualities of the Graeco-Roman Hermes-Mercury and the Egyptian deity Thoth.

Hermes is the divine messenger mediating between heaven and earth, the trickster god of the crossroads, patron of both merchants and thieves. Thoth, meanwhile, is the patron of the sacred sciences, also a mediator, understood as operating at every level of being. He serves the gods, but also precedes and even creates them, for he is the self-creating arch-magician, the Word in action. He has but to name a thing and it springs into life.

Appearing at the meeting point of history, legend, and myth, Hermes Trismegistus is a tricky character to pin down. He shifts roles and identities from one moment to the next. As the archetypal trickster he is the inner and outer teacher, the balancing point between all polarities, often referred to as Hermes or Mercurius.

Credited to Trismegistus is a body of writings called *The Hermetica*. Written down in early Christian Alexandria, but clearly of more ancient inspiration, and practically unknown in Europe until the Renaissance, their impact was considerable. Hermes describes man as *the great miracle*, the microcosm made "in the image of God," with all he needs to achieve his divine destiny. The best-known Hermetic text is the enigmatic guide to the Great Work known as *The Emerald Tablet* (*see opposite page 1*) whose meaning is inexhaustible.

Ibis-headed Thoth (above left), the divine scribe in the ancient Egyptian pantheon holds aloft the Ankh and two serpent-entwined staffs. Hermes Trismegistus (above right) holds aloft an Armillary sphere to indicate the Above, while pointing to the Below. Mercurius (below) mediates between antagonistic polarities, and holds the Caduceus symbol of harmony in each hand.

POTION MAKERS
physician, heal thyself!

One of the main subjects of the *Hermetica* is medicine, and in a series of texts Hermes instructs Asclepius, the semi-divine healer of Greek mythology. The Staff of Asclepius, a stick with a snake winding up it, is the international symbol for medicine, and Hermes' Caduceus, with two snakes, has also become widely used as a medical symbol. The perfect equilibrium symbolized by the Caduceus is the aim of all holistic medicine. Thus all alchemists, as "Sons and Daughters of Hermes" consider themselves healers, and often refer to the Stone itself as the Universal Medicine. Jabir Ibn Hayyan [721-815], Michael Maier [1568-1622], and Robert Fludd [1574-1670], were all notable physicians as well as legendary alchemists.

The first thing to be healed by the alchemist is the very thing from which a medicine is made. In making a medicine from rosemary, for example, the alchemist seeks to perfect the plant itself. While a chemist might consider the resulting potion simply a combination of purified compounds from a dead plant, to the alchemist it represents the very idea of rosemary. As such it is more alive than ever before, in perfect resonance with its ideal form.

In order to understand how the alchemist can entertain such an extraordinary notion it is necessary to go back to the dawn of Creation itself and establish the metaphysical principles upon which the alchemical philosophy is based.

17

CREATION
in the beginning

To the alchemist Creation is the Great Work of the One, a universe ensouled and inspired. Contrary to contemporary theories that suggest that matter gives birth to consciousness, the alchemist's view of Creation is metaphysical – spirit precedes matter. Thus the Great Work of the alchemist is to restore fallen matter to spirit.

The *Hermetica* describes a compelling vision of Creation. Hermes witnesses the painful sacrifice of divine unity, the rending of the Void, as the mysterious advent of the *Logos* (the Word) precipitates a smoky darkness, which condenses to a "watery substance," the *Prima Materia*. The Logos is the "Son of God," the creative principle, eidos, which seeds the chaotic waters, which in turn become the matrix of all forms. Thus the One, through reflection, becomes Two, giving rise to a third principle which, like Thoth, mediates and governs this polarity, allowing their fruitful union by acting both as generative spark and midwife.

Thus are established the three Philosophical Principles – Sulphur ♄ (Logos/eidos), Mercury ☿ (Prima Materia/*hyle*) and Salt ⊖. If a mercurial, almost paradoxical, factor can already be discerned in this scenario, this is to be expected and accepted. Nothing other than the Absolute makes absolute sense and the alchemist must be both supple and subtle in his or her understanding. These lofty concepts are brought down to earth in the alchemist's kitchen, as shall become clearer later on.

The first ideas to manifest from the interaction of the Three are the Four Elements – the template for all created things.

In the beginning. The creation
of heaven and earth.

The earth was without form;
darkness was upon the face of the deep.

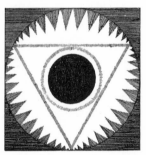

The Spirit moved upon
the waters' face.

The Command was given; and there
was light. And it was good:

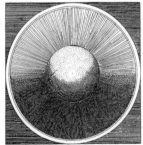

And the light, called Day, was divided
from the darkness, called Night.

Let there be a firmament in their midst
that divides the waters from the waters.

19

THE ELEMENTS
fire, water, earth, and air

The four Philosophical Elements are symbolized by triangles. Ascending Fire △ and Air ⊿ are upward pointing triangles, while descending Water ▽ and Earth ⊽ point downward. The triangles of Air ⊿ and Earth ⊽ are crossed, being relatively less ascending and descending respectively. The four Elements as a group are symbolized by the cross + (*page 58 lists all the symbols used in this book*). As archetypal forms preceding the manifestation of matter, these Elements are not to be confused with the atomic elements, nor with the common substances with which they share their names.

Each element shares its qualities with two others (*opposite*). This provides the dynamic that allows for cyclic transformation within matter, known as the *Rotation of the Elements*. △ is the most volatile of the Elements, ⊽ the most fixed. △ and ⊿ are the masculine Elements, ⊽ and ▽ feminine. Alchemists see all things as *mixta*, mixtures of the four Elemental qualities. For example, common water and alcohol are both Water, but alcohol ("firewater") has more elemental △ in it, while water has more ⊿.

In traditional cosmology, the first things created from the Elements are the Heavens, the Zodiac, the fixed stars and then the seven planets, literally "wanderers," each of which has specific qualities that have particular resonances with all things on Earth.

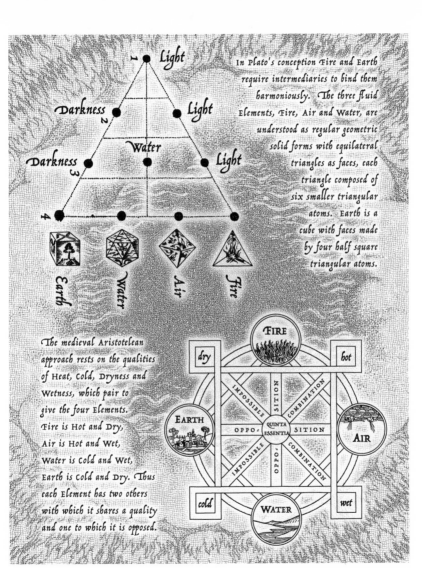

1 Light

Darkness 2   Light

Water

Darkness 3   Light

4

Earth   Water   Air   Fire

In Plato's conception Fire and Earth require intermediaries to bind them harmoniously. The three fluid Elements, Fire, Air and Water, are understood as regular geometric solid forms with equilateral triangles as faces, each triangle composed of six smaller triangular atoms. Earth is a cube with faces made by four half square triangular atoms.

The medieval Aristotelean approach rests on the qualities of Heat, Cold, Dryness and Wetness, which pair to give the four Elements. Fire is Hot and Dry, Air is Hot and Wet, Water is Cold and Wet, Earth is Cold and Dry. Thus each Element has two others with which it shares a quality and one to which it is opposed.

FIRE

dry   hot

IMPOSSITION   SITION   COMBINATION

EARTH   OPPOS   QUINTA ESSENTIA   SITION   AIR

IMPOSSIBLE   OPPO   COMBINATION

cold   wet

WATER

HEAVENLY METAL
the magnificent seven

To the alchemist, as to the ancients, the seven planets are the celestial forms of seven divine beings. The traditional planetary order is based on the planets' speed against the fixed stars and was eventually recorded by the Chaldaeans [c. 700 BC]. A remarkable pattern relates this order to the days of the week (*below*).

The brightest heavenly bodies are the Sun ☉, the Moon ☽, and Venus ♀, corresponding to gold, silver, and copper, the three native metals found shiny and workable on Earth. Ancient metallurgy unearthed four more pure metals; iron, tin, lead and quicksilver, corresponding to four remaining unpaired planets. Slow moving Saturn ♄ matched ponderous lead, fiery red Mars ♂ ressembled warlike iron, speedy Mercury ☿ echoed fluid quicksilver, and tin crackled like the thunderbolts of Jupiter ♃.

In seeking to refine and purify their own souls, alchemists also discover the planets, from base ♄ (Saturn/lead) to saintly ☉ (Sun/gold), resonating within themselves. We still use such adjectives as saturnine, mercurial and jovial to describe peronalities that reflect particular planetary qualities. Animals and plants also have planetary qualities – lions are solar, unicorns are lunar, while spiky plants are ruled by Mars, and apples by Venus.

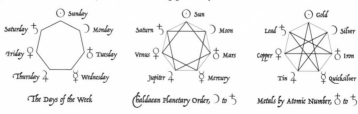

The Days of the Week   Chaldaean Planetary Order, ☽ to ♄   Metals by Atomic Number, ♂ to ♄

The planetary symbols use three components; Sun or Sol ☉, Moon or Luna ☽, and the four Elements ✛. Above, the work of the White begins at the bottom left with Saturn ♄, the lunar principle obscured in blackness by the elements, as it emerges it pairs with the elements as equal in Jupiter ♃, before freeing itself as resplendant Luna ☽.

On the right the work of the Red begins with colourful Venus ♀, the solar principle dominating but tied to the elements, in Mars ♂ the solar is subsumed by the elements only to then free itself as pure centralized Sol ☉. Mercury oversees the Work. As the very possibility of continuity, his glyph necessarily uses all three components ☿.

MINERALS AND PIGMENTS
the secret colors of the Art

The ancients found many practical ways of expressing planet-metal associations, particularly in sacred art. Amongst the earliest known pigments were the iron oxide ochres employed by palaeolithic cave painters [c. 300,000 BC onward]. Early metallurgists, potters, and glassmakers found more exquisite colors in naturally occuring metallic ores and minerals. More recently, the ancient Egyptians, masters of the art of creating magnificent colors, invented the peerless Egyptian blue (a copper silicate), the earliest known artificial pigment and the first to capture the color of the heavens (*see page 48*).

The Crucifixion (*opposite*) by Raphael [1483-1520] is a striking example of pigments being used in full awareness of alchemical planet-metal correspondences, even employing the same traditional planetary arrangement shown on the previous page.

The Sun ☉ and the Moon ☽ are rendered with gold and silver, while Venus ♀ and Mars ♂ provide the copper and iron pigments for the green robe of the angel below the Sun ☉. Saturn ♄ and Jupiter ♃, in turn, supply lead-tin yellow for the garment of the angel below the Moon ☽. In the middle, Christ's blood and loincloth are painted with the quicksilver-sulphur compound vermilion, the pigment color held by the Chinese to represent eternal life.

24

I·N·R·I

☉ Gold   ☽ Silver

malachite
verdigris &
green earth

red vermilion
or cinnabar

lead-tin
yellow

As Above
so below

However pigments and potions are mixed, they will not be truly alchemical unless they are made at the right moments. Timing is crucial to maximize planetary resonances, and this requires an understanding of the heavenly movements.

The seven wanderers move through the twelve constellations of the Zodiac that divide the solar year, their constantly changing positions determining a unique balance of qualities for each moment. Internally the planets represent seven specific modes of the soul that the alchemist must develop to progess in the Great Work, while the Zodiac corresponds to twelve processes that the soul must cyclically endure on the path of return to the Absolute.

In the northern hemisphere the astrological and alchemical year begins with Aries at the spring equinox, when day and night are of equal length. The process through spring to midsummer then marks the ascent of the Sun, which later declines toward its midwinter death and subsequent spring rebirth. Tied to this, the vegetable realm, as the most immediately solar-dependent kingdom, flourishes and recedes with the solar year, while the monthly waxing and waning of the Moon controls its juices, drawing the sap to the upper parts and back down to the roots. The herbal alchemist is therefore compelled to heed the injunctions of Paracelsus that he:

"… should know the innate nature of the Stars, their complexion and property, as well as a physician understands the nature of a patient, and also the concordance of the Stars, how they stand in relation to … all things that grow and spring from the matrices of the Elements. … Medicine is without value if it is not from Heaven."

ARIES = CALCINATION
action of fire on minerals in air

TAURUS = CONGELATION
thickening by cooling

GEMINI = FIXATION
trapping a volatile as a solid or liquid

CANCER = SOLUTION
dissolutions or reactions of substances

LEO = DIGESTION
prolonged continuous gentle warming

VIRGO = DISTILLATION
ascent and descent of a liquid

LIBRA = SUBLIMATION
ascent and descent of a solid

SCORPIO = SEPARATION
isolation of insoluble from soluble

SAGGITTARIUS = CERATION
softening hard material

CAPRICORN = FERMENTATION
biological animation of a substance

AQUARIUS = MULTIPLICATION
increasing the potency of the Stone

PISCES = PROJECTION
the mysterious action of the Stone

SPAGYRICS
putting the djinn in the bottle

And so to potions! Spagyria is a term coined by the great Germanic physician Paracelsus [1493-1541] from the Greek *spao* – to draw out; and *ageiro* – to gather. It is equivalent to the alchemical dictum "*Solve et coagula!*" (dissolve the fixed and fix the volatile), and has become a general term for the production of alchemical medicines. Making spagyric potions is the ideal way to start getting to grips with alchemical ideas and practice, and most of the key processes can be carried out in the kitchen with some basic equipment. When making a potion from a particular plant, work should begin on the day of the week and during the planetary hour that correspond to its planetary rulership (*see pages 56-58*).

For a basic spagyric tincture a herb is ground up and macerated (left to steep) in warm grape brandy inside a sealed jar for two weeks. The brandy, which already contains plant Mercury ☿ (alcohol) becomes infused with the herb's Sulphur 🜍 (essential oils). The tincture is then filtered and the soluble Salt ⊖ painstakingly extracted from the plant residue (*see page 30 for instructions*). This separates the inessential from the essential, the subtle from the gross. Finally the ⊖ is added to the 🜍 and ☿ tincture, recombining the three Principles. The only thing discarded is the insoluble plant residue.

Distillation (*opposite*) is key to more sophisticated spagyric work. A plant's 🜍 may be extracted by distilling it in water. The 🜍 collects on the surface of the distillate and is easily drawn off. ☿ can be extracted by fermenting the plant (*see page 55*), but since ☿ is universal, the same in all plants, any ethanol distilled to at least 96% purity will do.

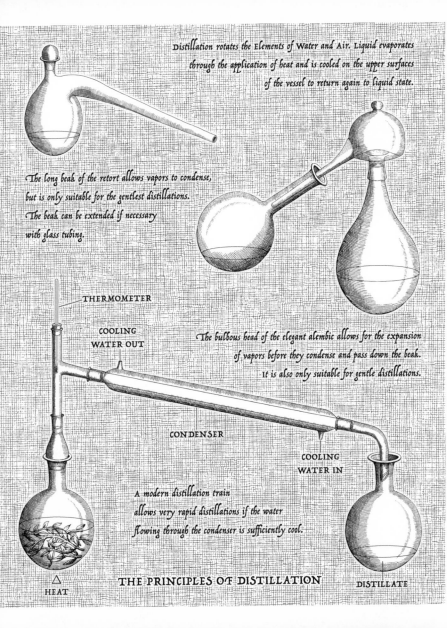

Distillation rotates the Elements of Water and Air. Liquid evaporates through the application of heat and is cooled on the upper surfaces of the vessel to return again to liquid state.

The long beak of the retort allows vapors to condense, but is only suitable for the gentlest distillations. The beak can be extended if necessary with glass tubing.

THERMOMETER

COOLING WATER OUT

The bulbous head of the elegant alembic allows for the expansion of vapors before they condense and pass down the beak. It is also only suitable for gentle distillations.

CONDENSER

COOLING WATER IN

A modern distillation train allows very rapid distillations if the water flowing through the condenser is sufficiently cool.

△ HEAT

THE PRINCIPLES OF DISTILLATION

DISTILLATE

ANGEL WATER
collecting the secret fire

Nature is full of secret bounty. Common dew is the distilled essence of Heaven and Earth, a condensation of the Universal Spirit; the *Secret Fire*. The best way to collect it is to use purified plant salts, which are highly *hygroscopic* and absorb dew from the air. Plant Salt $\ominus$ is understood alchemically as a transitional substance, since it bridges two kingdoms, in this case vegetable and mineral.

1. Burn any plant matter to ashes, oak bark is best. 2. In a large pot add the ashes to 20 times their volume of rainwater. 3. Boil for 20 minutes to extract the water-soluble $\ominus$. 4. Cool and filter into a large pan. 5. Evaporate the liquid, stirring rapidly as the $\ominus$ starts to solidify. 6. Grind the dry $\ominus$ and heat it in a pan. This is called calcination, literally "making like chalk." 7. Calcine for several hours at around $500°C$ – full blast on a gas stove. 8. Dissolve the cooled $\ominus$ in filtered rainwater. 9. Repeat stages 4 to 7 at least twice until the $\ominus$ is really white. 10. Repeat stages 1 to 9 until you have at least two cups of $\ominus$. 11. In the late evening, ideally on a fine spring night during the waxing $\supset$, spread the $\ominus$ out thinly in flat glass or porcelain dishes. 12. Place the dishes in an open spot outside, raised well off the ground. 13. At sunrise collect the dishes and pour their contents into a distillation flask, avoiding all contact with skin or metal. The $\ominus$ should have liquified, at least partially. 14. Gently distill off this "Angel Water" until the $\ominus$ are dry. 15. Pour into a dark glass jar and seal tightly. 16. Save the $\ominus$ likewise, which can be used countless times for this purpose, becoming increasingly charged with Secret Fire.

The $\ominus$ made in this way is the *Sal Salis* (Salt of the Salt), the Salt proper, but there is another $\ominus$ called the *Sal Sulphuris* (Salt of the Sulphur), which is extracted from the plant soup remaining after ♀ or ☿ have been distilled. This is boiled down to a tar, burned, ground, reduced to ashes and then extracted like the Sal Salis.

Angel Water can be used as a tonic (a few drops in water for a bright eye and a shiny coat), saved for use in other potions, or it may be further developed by making the elegant *Archaeus of Water*.

It ascends from Earth to Heaven and descends again to Earth, Combining the power of above and below.

ARCHAEUS OF WATER
fractional distillation

To master the art of distillation takes considerable care and experience. Alchemists such as Hieronymus Brunschwygk [1450-1513] and John French [1616-1657] devoted great tomes to the subject. Distillation is a rotation of Elements – a fluid is heated to evaporation point, becoming a gas that recondenses back to fluid upon contact with a cool surface. To assemble a distillation train, procure a borosilicate glass distillation flask, a simple condenser and a glass receiving vessel (*see page 29*). For a rapid distillation, direct heat can be applied to the distillation flask; for more gentle distillations place the flask in a water bath, heating from below; for hotter distillations at a constant temperature use an ash or sand bath.

The alchemist knows many types of Water – elemental Water, Chaotic Water (hyle), and various other substances, mysteriously described as "Our Water." Even common water is not just one thing, it is a fluid of subtle variety, the only liquid that expands upon freezing, with vital magnetic and mediating properties.

Angel Water can be used to prepare a potion known as the *Archaeus of Water*, the method for which uses *fractional distillation* to separate water into twelve Philosophical parts (*shown opposite*). A few drops will enliven any other water used for purposes such as fermentation. Each of the twelve waters, prior to their recombination, is suited to different purposes – for example through repeated distillations one water can be made sharp enough to act on metals.

Distilling volatile fluids can be explosively dangerous. Many an alchemist's kitchen has been reduced to ashes. Protect yourself!

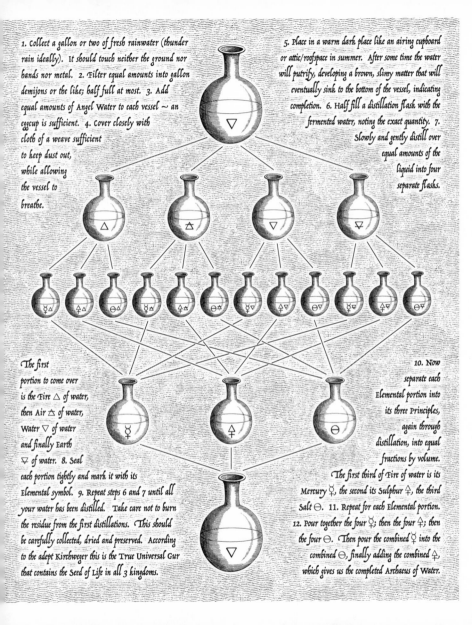

1. Collect a gallon or two of fresh rainwater (thunder rain ideally). It should touch neither the ground nor hands nor metal. 2. Filter equal amounts into gallon demijons or the like; half full at most. 3. Add equal amounts of Angel Water to each vessel ~ an eggcup is sufficient. 4. Cover closely with cloth of a weave sufficient to keep dust out, while allowing the vessel to breathe.

5. Place in a warm dark place like an airing cupboard or attic/roofspace in summer. After some time the water will putrify, developing a brown, slimy matter that will eventually sink to the bottom of the vessel, indicating completion. 6. Half fill a distillation flask with the fermented water, noting the exact quantity. 7. Slowly and gently distill over equal amounts of the liquid into four separate flasks.

The first portion to come over is the Fire △ of water, then Air ◬ of water, Water ▽ of water and finally Earth ▽ of water. 8. Seal each portion tightly and mark it with its Elemental symbol. 9. Repeat steps 6 and 7 until all your water has been distilled. Take care not to burn the residue from the first distillations. This should be carefully collected, dried and preserved. According to the adept Kirchweger this is the True Universal Gur that contains the Seed of Life in all 3 kingdoms.

10. Now separate each Elemental portion into its three Principles, again through distillation, into equal fractions by volume. The first third of Fire of water is its Mercury ☿, the second its Sulphur ♄, the third Salt ⊖. 11. Repeat for each Elemental portion. 12. Pour together the four ☿; then the four ♄; then the four ⊖. Then pour the combined ☿ into the combined ⊖, finally adding the combined ♄, which gives us the completed Archaeus of Water.

PRIMUM ENS
salt volatilization

Having mastered the art of distillation, and the extraction of salts, the potion-maker may be ready to attempt a superlative spagyric potion, much revered by Paracelsus, and called by him the *Primum Ens* (First Being). The profound integration of ♁, ☿, and ⊖ acheived by this process raises a plant to the same resonance as its spiritual blueprint, maximizing its healing potential.

> *Materia: 1. Pure plant Mercury ☿ (ethanol), made by carefully distilling brandy seven or eight times (Spirit of Wine), or alternatively bought commercially (ideally made from grapes). 2. Plant Sulphur ♁ (essential oil, for example rosemary) – we can extract this ourselves (see page 28) or it can also be bought from a good source. 3. Salt ⊖ from the same plant (see page 30). Method: 1. Decant 150 ml of ♁ into a 500 ml retort with a vent at the top. 2. Little by little, via the vent, add 30g of pure dry ⊖ from the same plant as the ♁. 3. Gradually heat the retort in a sand bath to a very gentle simmer, so that it distills over gently into a flask. You should notice after a while a delicate "snowfall" of tiny particles over the simmering oil. This gradually increases, rising to the throat of the retort and frosting the glass. This is a spagyric wonder, the Volatilization of Salt. 4. When the residue turns to a honey-like consistency halt the distillation. 5. Return the ♁ to the retort and distill again. This time the ♁ will wash the ⊖ down into the receiving flask. 6. Distill again and the ⊖ will again frost the throat of the retort. 7. Clean the retort with turpentine and allow to dry. 8. Distill again, adding 150 ml pure plant ☿. All the ⊖ will come over, combined with the ♁ and ☿.*

This recipe, like many others in alchemy, can defeat even the experienced chymist (hermetic chemist), unless he or she attends closely to every single part of the process – traditionally, secret recipes were deliberately confused to foil the unworthy. A teacher can help, but these days adepts are thin on the ground. However, if you do find yourself becoming frustrated, remember the alchemical adage "when the student is ready, the master will appear."

Michael Maier, Symbola Aurea Mensae, 1617

Avicenna (Ibn Sina) demonstrates the requirement to volatilize (eagle) the fixed (toad).

Michael Maier, Atalanta Fugiens, 1618

The Celestial Salt containing the Secret Fire above and the Salt of the Earth below.

CIRCULATUM MINUS
the lesser work

The *Circulatum Minus* represents the culmination of plant alchemy and is a very tricky potion indeed, mastered by only a handful of people since the method was first published in London by Baron Urbigerus in 1690 (*a summary of the method is given on the right*). Its name means Lesser Circulation, the Greater Circulation being the Philosopher's Stone itself (circulation is simply a gentle distillation in a closed vessel, achieved when the temperature inside the vessel is just enough for a continuous evaporation and recondensation).

The Lesser Circulation actually involves digestion and distillation rather than a circulation, suggesting instead a mysterious exaltation of the materia similar to the Stone. Great patience is required, and purity is of the essence; the matter must remain uncontaminated.

If successful the Circulatum should have a peculiar penetrating odor and a sharp corrosive taste. The test is as follows: Cut up fresh green leaves from an aromatic herb like mint and immerse them in the matter. The liquid will cloud as tiny drops of oil form and rise to the surface. Eventually the exhausted dregs fall to the bottom. The oil contains the combined Principles of the plant. This oil can be separated and the remaining Circulatum redistilled from the vessel and stored for future use.

Whoever masters this process can truly be called an alchemist.

Materia: The purified Principles of a plant ~ $\ominus$, φ and $\math815{Y}$ (Melissa (lemon balm) works well but has very little φ); Canada or co-
paiva balsam.   Method: 1. Gradually imbibe the $\ominus$ with equal quantities of φ and balsam until it is loosened just to the point of wetness.
2. Add a little bit more balsam for luck.   3. Digest at about 40°C in a glass jar, covered so it can still breathe.   4. Stir 9 or 10 times a
day with a wooden spatula, adding more balsam when necessary to maintain the same consistency.   The $\ominus$ should be fully saturated after
about four weeks and have dissolved to a dark, glassy honey.   5. Add six to eight times the volume of pure $\math810{Y}$.   6. Seal and digest at 40°C for
at least ten days, stirring several times daily.   7. When a change of color has been observed and the $\ominus$ takes on a slimy appearance, distill
gently in a water bath, taking care that only the $\math815{Y}$ comes over, not the balsam.   8. Cohobate (i.e. return the distillate to the flask) and
redistill as before for a total of seven times.   9. Distill one last time and you may have the Lesser Work of alchemy.

FROM MINOR TO MAJOR
transformation to transmutation

Having reached the pinnacle of the plant work, the alchemist is ready to proceed further. While the Circulatum Minus effects an apparently miraculous *transformation*, the Greater Circulation is said to go one step beyond, actually *transmuting* elemental metals into gold.

The accounts of respected, and previously sceptical, authorities, such as scientist Van Helmont [1580-1644] and 17th century physician Helvetius, describe their incredulity as they witnessed and then personally performed transmutations of base metal into gold using a *powder of projection*, produced by mysterious strangers who sought them out and then vanished. In the case of the legendary French alchemist Nicolas Flamel [1330-1418], it was the chance purchase of a strange and ancient book that ultimately lead to his discovery of the Stone, and the attainment of fabulous wealth and immortality.

But how can such things be possible? The enigmatic adept Fulcanelli may shed some light on the process, as quoted by Jacques Bergier, a French nuclear physicist who met him in 1937:

> *"There is a way of manipulating matter and energy so as to create what modern science calls a force-field. This force-field acts upon the observer and puts him in a privileged position in relation to the universe. From this privileged position, he has access to realities which are normally concealed from us by time and space, matter and energy. This is what we call the Great Work."*

Common language cannot adequately descibe such things. To approach an understanding we must identify entirely with the alchemical perspective, while in order to move from the speculative to the operative we must also identify entirely with the materia itself. In alchemy the empathic participation of the alchemist is key.

The Alchemist as Master of Polarities (above), and praying for illumination in his lab-oratory (below).

OPUS MAGNUM
paradise regained

The goal of the Great Work is nothing less than union with the Absolute. Before this process can even begin, however, the lower reconciliation of spirit and soul must take place, requiring the total capitulation of the lower self. The Work begins when the traveler has reached the end of his tether, and realizes that there can be no further progress away from the Source. The alchemist is on his own.

Freed from association with a false identity, spirit and soul can embrace. The purified Sulphur and Mercury must now marry and give issue to the hermaphroditic child, a process known as the "Work of the Sun." Hermetically sealed in a glass egg from now on, the crucial matter, freed from all inessentials, is incubated in seclusion (*symbolized by the first flask in the emblem series opposite*).

The union of soul and spirit results in the conception of a new being embodying both principles, which is then subsumed by the appearance of the Crow's Head, signalling the *Nigredo*, the awful black phase when all seems lost. Hopefully the matter begins to lighten, but then a separation takes place and all appears to volatilize, rising and falling like goose down. From these ashes springs new life; the three flowers in the bottom left flask symbolizing the purified Tria Prima – the Body too has been resurrected.

It all sounds so easy, but the majority of alchemists never succeed in even reaching this point, having started with the wrong material.

CONCEP - TIO.

PRÆG - NATIO.

COLOR COELESTINUS.

COLOR COELESTINUS.
cum tua terra nigra.

Union of the
purified Principles.

The fixed is volatilized,
female absorbs male.

Female and male unify.
A heavenly blue appears.

The black earth appears
within the blue.

Caput
Putre -
Philo -

Corvi
factio
sophorum.

Caput
et lac
dealba

Corvi
Virginis
tur

Caput
Separatio
à

Corvi
animæ
corpore.

Caput
totalis
animæ

Corvi
separatio
à corpore.

The crow's head,
philosophic putrefaction.

The crow's head, whitened
by the Vigin's milk.

The crow's head, soul
separating from body.

The crow's head,
complete separation.

Cinis
Cinerem
vili

Cinerum,
hunc ne
pendas.

Medicina
Eli -

alba sive
xir album.

Medicina
Elixir

Rubea sive
rubeum.

Projectio

Augmenta
tiog.

The remaining ash
is not to be despised.

The White Elixir,
first degree of perfection.

The Red Elixir,
perfect fixity.

Projection augments
the power of the Stone.

J. D. Mylius, Anatomia Auri, 1628

41

LAPIS PHILOSOPHORUM
the Philosopher's Stone

Key to the elaboration of the Stone is the *Prima Materia*, whose identity is the greatest secret of alchemy. This primordial matter is in all created things, but there is only one substance from which it can be drawn out and purified. What is this One Thing or *primum agens*? In the ancient texts the adepts answer only in riddles. "It is a stone that is not a stone," "... it is thrown into the street by servant maids, children play with it, yet no one prizes it..." As common as dirt, it is everywhere to be found and is everywhere "esteemed the vilest and meanest of earthly things." If this thing is identified the Prima Materia can be released from its fetters, the subtle separated from the gross, whereupon the Philosophic Mercury comes forth – and the rest of the process is "women's work and child's play."

The Work begins with the union of the liberated Principles (*previous page*). By applying heat "gently and with great ingenuity" we allow Nature to take its course. The progress of the Work is observed according to the colors that the matter displays. If the *Nigredo* is survived there should be a yellow dawn followed by a peacock's display of colors that heralds the white *Albedo* stage, the arrival of the White Queen, the White Lion, or the swan, the Elixir that transmutes to silver and confers immortality if ingested. The subsequent reddening of the matter is the *Rubedo*, the triumphal arrival of the Red King, the phoenix. This is the Philosopher's Stone, a weighty, glistening, waxy powder that acts as a universal medicine in all three kingdoms and, when digested with gold, becomes the Powder of Projection that transmutes a thousandfold its weight of molten metal into gold.

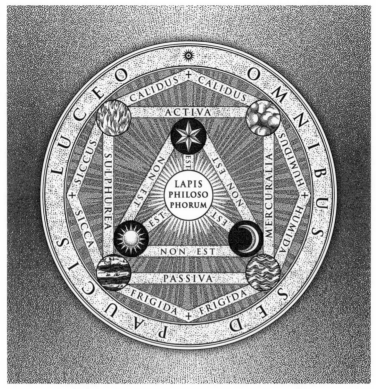

THE PHILOSOPHERS' STONE ~ *"For all, but in few I shine."*

THE MACROCOSM

MAN, THE MICROCOSM

THE DIVINE

43

GOING FOR GOLD
onwards and upwards

This book can only serve as an introduction to alchemy. Making alchemical potions requires a solid grounding in herbalism and astrology, as well as a good grasp of alchemical principles. Actually using such potions responsibly and effectively is a different matter. Many spagyrists are not physicians and leave the prescribing of their remedies to the discretion of the medical practitioners they supply.

An effective way to use spagyrics as part of one's own process is to make a potion for each day of the week using safe herbs corresponding to the planet in question (*see page 56*). A couple of drops of a Sun potion on a Sunday, a Moon potion on a Monday and so on helps tone and harmonize the inner cosmos.

If your appetite has been whetted, and you find yourself rolling up your sleeves in excitement, do be prepared for many frustrations and even disasters along the way. Mercurius, as we have warned, can be a very tricky guide, punishing complacency, overeagerness, and carelessness. For those who proceed with sufficient care, however, the rewards are incomparable. If you let yourself be guided by the alchemical dictum "*ora, lege, lege, lege, relege, et labora!*" – pray, read, read, read, reread, and work! – the glory of the world may be yours.

"Between eternal birth, resurrection from the fall and the discovery of the Philosopher's Stone, there is no difference."   Jacob Boehme, shoemaker and mystic. [1575–1624].

45

APPENDIXES - BASIC METALLURGY

From simple beginnings in around 6000 BC to the end of the thirteenth century metallurgy knew only the seven metals of antiquity; gold, silver, copper, iron, tin, lead and quicksilver.

☉ **GOLD** (melting point 1064°C) is highly malleable, and is thus easily worked. Gold is found in native form in deposits that can be mined. Where these deposits have eroded gravelly stream beds can be found bearing preciously rare gold nuggets. Early gold artefacts often have silver "impurities" and the ancients named the alloy of gold and silver **ELECTRUM**. Gold can be separated from the silver using **CEMENTATION** — mixing the gold-silver alloy with common salt allowed the silver to form a soluble chloride that washed away. Aqua fortis (*see page 50*) can also be used to separate silver from gold as it only dissolves the former.

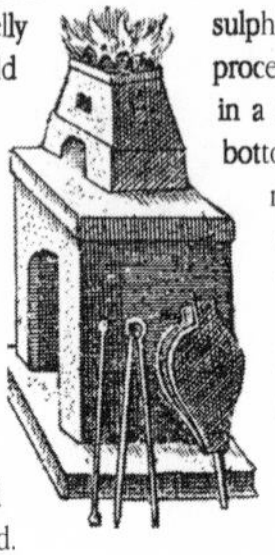

☽ **SILVER** (melting point 962°C) is second only to gold in ductility and malleability, and like gold is easily worked. Silver occurs in native form, but very rarely. It will tarnish and blacken when exposed to sulphur or hydrogen sulphide in the air. The lead ore galena always contains some silver. If the lead is heated to ash, forming lead oxide, a small bead of silver remains. If a crucible made from bone ash is used it will absorb the lead oxide. This process, known as **CUPPELLATION**, was the main means of silver production for millennia.

♀ **COPPER** (melting point 1085°C) is malleable and ductile. It was the first metal to be widely made into weapons and tools, from around 6000 BC onward. Copper was first worked in a similar way to stones, but if hammered repeatedly it becomes brittle. This can be remedied by **ANNEALING**, prolonged heating sufficient to make the metal glow, followed by slow cooling. The earliest smelted copper artefacts appeared around 4000 BC. Green malachite was the copper ore used by early smelters. It is possible that malachite placed in potters kilns, at temperatures of 1100-1200°C, and subsequently forming copper nodules was the initial inspiration for smelting.

♂ **IRON** (melting point 1538°C) is the most common metal on Earth, but is almost never found in native state. It was found by the ancients in the form of meteors, which were initially worked like stone. Manmade iron may have been available as early as around 2500 BC but was not common until over a millennium later. Iron ore is easily **REDUCED** by charcoal but can only

be forged if produced at temperatures of over 1100°C. Early smelted iron was a spongy mass mixed with waste slag that was heated and hammered to expel the slag and then forged.

♄ **LEAD** (melting point 327°C) is highly ductile and malleable and does not corrode easily. It is not found in native form but its sulphide ore galena appears very metallic. Galena can easily be processed to give pure lead, and this could have been achieved in a camp fire, in which molten lead would collect at the bottom. This is a key to a central metallurgical process - the reduction of an ore must be accompanied by a change to a liquid that allows the metal to flow and separate from the still solid waste material, or *gangue*.

♃ **TIN** (melting point 232°C) is not found in native form. It is malleable and ductile and quite resistant to corrosion. Tin objects date from around 2000 BC, and it was smelted by reduction with charcoal. Initially tin was thought to be a form of lead. Early copper smelters discovered that mixing different ores produced an easier flowing stronger metal - this metal was **BRONZE**. Tin's highly crystalline structure means that when deformed it gives an audible cry.

☿ **QUICKSILVER**, or mercury, (melting point -39°C) is the only metal that is liquid at room temperature. Early purification techniques included squeezing it through leather. It is highly poisonous and has long been known as such. Extraction of quicksilver from ores such as cinnabar (mercuric-sulphide) is carried out by distillation since mercury compounds decompose and volatilize at moderate temperatures. Quicksilver will dissolve gold and silver, and this process of **AMALGAMATION** was often used to separate these metals from impurities.

Four other metals were discovered in the middle ages. **ARSENIC** was discovered by Albertus Magnus (1193-1280) when he heated arsenious oxide with twice its weight of soap. **ANTIMONY** was formed by roasting stibium, antimony sulphide, in an iron pot. **BISMUTH** was produced at the end of the 16th century by reducing the oxide with charcoal. **ZINC** was known in China around 1400. It too was reduced from its oxide using charcoal. In the late 18th C zinc was added to liquid copper to make the first **BRASS**. In the new world **PLATINUM** was used by pre-Columbian Native Americans, only becoming known to Europeans in the sixteenth century.

CERAMICS AND GLASS

CLAY is a plastic (shapeable) substance formed by the gradual erosion of minerals such as feldspar to give minute particles of *hydrated silicates* mixed with water and other compounds. The discovery around 10,000 BC that clay heated to high temperatures changes to become hardwearing and strong is one of the turning points in mankind's story. Fired in a kiln at temperatures between 800°C and 1200°C, clay remains slightly porous and is known as *earthenware*. Firing at higher temperatures causes the clay to partially vitrify, producing *stoneware*. Porcelain is a fine white body fired to vitrification to become translucent. *Pyrometric cones*, which melt at different degrees of heat absorbtion, are often used to measure firing cycles and digital thermometers are also popular. It is also possible to judge kiln temperature by the color of the glowing ceramic (the temperature of metals can also be estimated from this list).

Lowest visible red to dark red	470-650°C
Dark red to cherry red	650-750°C
Cherry red to bright cherry red	750-800°C
Bright cherry red to orange	800-900°C
Orange to yellow	900-1100°C
Yellow to light yellow	1100-1300°C

Clay objects can be formed by hand and with simple tools, by turning on a potter's wheel or by pouring *slip* (a mixture of clay and water) into a mould. Once formed the clay is allowed to dry, after which it is known as *greenware* and is very brittle. It can be fired unglazed in a "biscuit firing," and then fired a second time with an application of glaze, or it can be fired in one cycle with or without an application of dry glaze.

Glaze, when fired, melts to form a hard glassy coating. It allows earthenware vessels to hold liquids. Mixed from finely ground ingedients it can be applied by dusting onto the clay object, or by mixing with water to be painted, poured on, or dipped into. Glazes are a specialized type of glass combining silica, alumina to increase viscosity when melted, and a flux to lower the melting point. Lead ♃ glazes use lead ♃ oxide as a flux. Soda ash, potash or other alkaline fluxes make alkaline glazes which often form crazing patterns of fine cracks as they cool. Opacifiers, such as tin ♃ oxide, are also used. Coloring materials are mixed with the glaze itself, applied on the clay body before glazing (underglaze) or on top of the glaze (overglaze). Examples include; iron ♂ oxide for ambers and browns, copper ♀ oxide for greens and turquoises, cobalt oxide for blues, and manganese dioxide for lilac, purple and brown.

GLASS is a strong, hard wearing, inert, biologically inactive, and of course transparent material. It is made primarily from silica (*silicon dioxide*), the most abundant mineral on Earth. Normal solids have regular molecular structures, however, many materials if cooled quickly assume a non-crystalline solid structure - a glass in the general sense. Silica is one of the few materials that forms a glass at normal cooling rates. Pure silica has a melting point of 1723°C. To reduce this melting point to about 1000°C soda ash or potash is added, and lime is added to counter the solubility that soda ash or potash cause in the glass. The mix is then heated in a kiln at about 1100°C until fused. Other ingredients sometimes used are lead ♃, which imparts more brilliance, and boron that improves the thermal properties useful for labware.

Glass normally has a green tinge from iron ♂ impurities, but an entire rainbow of colors can be made using different metals. Metallic gold ☉ in small concentrations produces a ruby glass. Silver ☽ compounds produce colors from orange-red to yellow. Adding more iron makes a stronger green. Copper ♀ oxide produces a turquoise color, while metallic copper produces a very dark opaque red. Cobalt makes blue glass. Manganese can be added for an amethyst color. Tin ♃ oxide together with antimony and arsenic oxides makes opaque white glass.

Glass was first manufactured around 2500 BC. The Ancient Egyptians made small jars and bottles by winding continuously heated glass threads around a bag of sand on a rod. Glass blowing was discovered in the first millennium BC and enabled the quick production of large leakproof vessels. Glassblowing uses three furnaces - one for the molten glass, a second for reheating the piece being worked on as necessary, and a third for annealing, i.e., cooling the glass slowly enough to avoid cracking and reduce stresses. As well as the blowpipe, tools used in glassblowing include shaping blocks, an iron rod known as a *ponty*, flat paddles, tweezers, and various shears. Intricate glassware ideal for an alchemical lab can be formed by heating, manipulating, and joining preformed rods, tubes and simple blown vessels using alcohol lamps, or nowadays propane or oxygen flames.

ARTISTS' PIGMENTS

PIGMENTS must be insoluble and reasonably light-fast. They are prepared for use by grinding finely into a paste with a little water using a glass muller on a glass surface (if coarse, grind in a pestle and mortar first). For oil paint use oil instead of water.

☉ **GOLD** hammered very thinly to make gold leaf can be applied to most surfaces. It can be made from leaf into a paint with gum arabic or gelatine, often called shell gold. ☽ **SILVER** like gold can be applied as leaf or made into a paint but it tarnishes in time with exposure to air.

♀ **COPPER** ores malachite (green) and azurite (blue) make good pigments, however the finer they are ground the paler their color. Pouring together strong solutions of blue vitriol (*copper sulphate*) and soda ash (*sodium carbonate*) precipitates an artificial malachite. Verdigris is *copper acetate*, soluble in water or alcohol for use, it can also be dissolved in resin in which case it will turn brown in air if not varnished. It can be grown as a crust on copper strips suspended in a mason jar with vinegar at the bottom, left in a warm place. The oldest known artificial pigment is Egyptian blue, a *copper silicate*. By dry weight: mix 10 parts limestone (whiting) with 11 parts malachite and 24 parts quartz. Grind thoroughly to homogenize. Add a flux of soda ash or potash, heat to around 900°C then keep at 800°C for at least 10 hours. Cool and grind for pigment. ♂ **IRON**. Red ochre, yellow ochre, raw sienna and raw umber are all *iron oxides*, the latter two are also well known in "burnt" forms made by calcining the raw. Natural green earth pigments contain *iron silicate*. Manmade *iron oxides* are also useful pigments ranging from yellows to reds to browns.

☿ **QUICKSILVER** in its red ore cinnabar, *mercuric sulphide*, makes a fine pigment. Vermilion is artificial cinnabar made by mixing together molten sulphur and quicksilver to form black *mercuric sulphide*. Heated in a suitable closed earthenware vessel this sublimates to form red *mercuric sulphide*, chemically the same compound, but transformed in color. Do not try this at home, quicksilver is highly toxic.

♄ **LEAD** pigments are toxic. Minium is *lead oxide*, a bright reddish orange, made by prolonged high temperature heating of lead in air. **WHITE LEAD** is *lead carbonate*. Place lead strips in earthenware jars with a little wine vinegar and digest somewhere warm. After some months a crust of white lead

should have formed. ♃ **TIN**. Now rarely used, lead-tin yellow (*lead stannate*) ranges from a light lemon yellow to a more pinkish color. Mix 3 parts minium thoroughly with 1 part tin oxide. Pass through a very fine mesh to help homogenize the mix. Heat slowly to 600°C, keep at this tempertaure for 2 hours, heat further, and keep at 800°C for another hour. Cool slowly.

COBALT is the key ingredient in smalt, a blue glass powder. Heat quartz and a potash flux with enough *cobalt oxide* to make an opaque blue glass to 1150°C to fuse. Remove while hot and plunge into cold water to break up before grinding into pigment. Cobalt blue, discovered in 1802, is *cobalt aluminate*. Grind 1 part *cobalt chloride* and 5 parts *aluminium chloride* and heat for 5 minutes in a test tube over a strong gas flame. **ANTIMONY** is used in Naples yellow, an artificial *lead antimonate* that dates back to Ancient Egypt, made by calcining a lead compound with an antimony compound. **ULTRAMARINE** is prepared from mineral lapis lazuli. Sprinkle finely ground lapis lazuli with linseed oil. Make a paste from equal parts of carnauba wax, pine resin and colophony. Add one sixteenth part linseed oil, one quarter part turpentine and the same of mastic. Mix 4 parts of this paste with 1 of the lapis lazuli and digest for a month. Knead the mixture in warm water until the blue particles separate and settle. Ultramarine was first synthesized in 1828.

LAKE pigments are made from organic sources such as madder (red), unripe buckthorn berries (yellow), ripe buckthorn berries (green) and cochineal beetles (carmen). Mix a saturated solution of potash and alum and mash the source matter in it until no more color comes out. Mix 6 spoons of alum with half a pint of warm water for each pint of colored potash solution. Pour in the alum solution to precipitate pigment. Insoluble **INDIGO** powder can be used as a pigment. The Maya made a fine artificial blue by heating a mix of indigo and *palygorskite clay*. 200°C for about 5 hours is suitable.

BONE BLACK. Boil animal bones (chicken bones are good) until fat free. Wrap tightly in aluminium foil and heat the package in a strong gas flame for an hour. Cool, unwrap and grind for pigment. **LAMP BLACK** is carbon gathered by placing a metal surface over a lamp flame. Not suited for painting its fineness makes it ideal for ink.

ARTISTS' MEDIA

PAINT is a mixture of pigment and a binder. **GUM ARABIC** is a popular water based binder, used to make **WATERCOLORS**, or, with an opacifier such as chalk added, **GOUACHE**. Crush pieces of gum arabic to a fine powder, add twice their volume of hot water, stir to dissolve. To reduce brittleness add a small amount of candy sugar. Mix 1 part gum arabic soultion with 2 parts pigment paste in water (all parts are by volume).

EGG TEMPERA is a very long lasting medium. Gently separate the white from the yolk then roll the yolk from palm to palm until dry. Hold the yolk membrane downward and pinch to release the liquid into a vessel while holding the membrane. Mix this yolk with equal parts of water or white wine vinegar, for use with pigment paste. **GLAIR** is made with egg white and is ideal for delicate illumination work on parchment. Beat egg white until the foam is dry. The liquid at the bottom of the vessel is glair.

SIZE is any coating that fills or coats a surface to protect and prepare it for the next layer. **RABBIT SKIN GLUE**, a specific type of animal glue (*see page 52*), is an excellent size and can also be used as a quick drying medium mixed directly with water based pigment paste. Soak 1 part rabbit skin glue in 18 parts water until swollen, then heat gently (without boiling) in a double boiler until dissolved. **CASEIN**, derived from milk, is a size that can also be used as a quick drying tough paint medium. Sift 2 parts powdered casein into 8 parts water and remove lumps by stirring. Add 1 part *ammonium carbonate* and allow to stand for half an hour, then add 8 parts of water. **STARCH** is another size: stir 1 part starch powder into 3 parts cold water to form a paste then slowly stir into 3 parts boiling water. When the solution starts to clear remove it from the heat. To use, dilute with water. **FISH GLUE**, extracted from fish by heating the skin or bones in water, is a good size for use on parchment.

OIL PAINT is surprisingly easy to make. Follow the instructions opposite for grinding pigment, using **LINSEED OIL**, **WALNUT OIL**, or **POPPY OIL**, and it is ready for use. If the pigments are already finely ground one can work them into paint with oil simply using a palette knife. Oil paints do not dry but harden through chemical reaction, giving oil painters time to work and rework their paintings. Ochres speed the oil's "drying," charcoal black slows it down. The craft of oil painting rests on

mastering the use of the many resins and spirits available. The following recipes are just a taste of the possibilities. Utmost care must be taken using volatile and flammable materials.

VARNISHES protect oil paintings, even if a glossy finish is not desired it is still best to apply one (after the painting has completely dried), and then use a wax finish. **GLAZES**, often very similar to varnishes, are used to thin paint to apply a veil of color while keeping it strong enough to adhere. The traditional rule is to paint "fat over lean," each successive layer having less pigment and more oils and resins. **DAMAR RESIN** comes in pale yellow lumps. Mixing equal parts in **PURE GUM SPIRITS OF TURPENTINE** (hereafter just "turpentine") and agitating daily until the resin is dissolved makes a good varnish that doubles as a glaze.

MASTIC makes a good varnish mixed and heated with twice its volume of turpentine. For a thin high gloss varnish mix 1 part **VENICE TURPENTINE** or **CANADA BALSAM** with 2 of turpentine. A good sweet smelling varnish for sized and sealed wood is 3 parts venice turpentine and 1 part **OIL OF SPIKE LAVENDER** warmed together to blend. **AMBER** varnishes are hard and versatile, serving as painting media, final varnishes, or as fixatives when thinned. Nowadays they are made with **COPAL** as a substitute: crush 1 part of the copal resin to powder and bottle with 4 parts benzene until nearly dissolved, then mix with 3 parts turpentine and heat gently to fully dissolve. If this warm solution is left unsealed the benzene will evaporate to leave a copal and turpentine varnish. A good wax finish to reduce gloss on a varnished painting is 1 part beeswax to 3 parts. **FIXATIVES**, used to fix pigment in place, are essential to the preservation of drawings in charcoal, chalk and pastel. Mix 1 part **SHELLAC** with 50 parts methyl alcohol. To use place the picture on the floor and blow the fixative vapor just above it to give an even coat.

ENCAUSTIC uses beeswax as a medium. Carefully heat 1 part beeswax and 3 parts turpentine until the wax has melted, then stir while cooling. Grind pigments thoroughly into the wax before applying with a brush or palette knife. Another recipe uses equal parts of soft elemi resin, beeswax, oil of spike lavender and turpentine. **FRESCO** is the specialized technique of painting directly on lime plaster. Mix lime-proof pigments to paste with water and paint directly onto fresh plaster.

ALCHEMICAL CHEMISTRY

MINERAL ACIDS are inorganic acids derived from chemical reactions. The discoveries of all three acids below are attributed to Jabir Ibn Hayyan (8th C.), as is Aqua Regia. NEVER add water to an acid, the reaction can generate considerable heat, and may boil and spit dangerously, ALWAYS add acids carefully to water.

OIL OF VITRIOL from the name *zayt al-zaj* given it by Jabir Ibn Hayyan, is better known to us as **SULPHURIC ACID**. It was originally obtained by dry distillation of green vitriol (*hydrous iron sulphate*) and blue vitriol (*hydrous copper sulphate*). When heated these decompose to oxides, giving off water and *sulphur trioxide*, which in turn combine as a dilute solution of sulphuric acid. The alchemist Johann Glauber (17th C.) prepared it by burning sulphur with saltpeter in the presence of steam. The saltpeter decomposes and oxidizes the sulphur to give *sulphur trioxide*, which combines with water to produce sulphuric acid. A later refinement of this process heats iron pyrite in air to form *anhydrous iron sulphate*. Heating to 480°C, this breaks down to give *iron oxide* and *sulphur trioxide* gas. This gas can be passed through water to produce sulphuric acid of any desired concentration. The diagram shows two ways in which one can prevent the water "sucking back" as this gas dissolves.

NITRIC ACID, originally known as **AQUA FORTIS** (strong water) or **SPIRIT OF NITRE** was first prepared by distilling green vitriol (*iron sulphate*) with saltpeter and alum. The following method is based on the one used by Glauber: Approximately equal parts by weight of concentrated oil of vitriol and saltpeter are placed in a retort. Heat the retort and brownish red fumes appear which can be condensed in a cooled reciever to a brown liquid. Further purification by distillations reduces the coloration from impurities to give a fuming aqua fortis.

SPIRIT OF SALT, or **HYDROCHLORIC ACID**, is made by reacting common salt with oil of vitriol (*sulphuric acid*) producing **SAL MIRABILIS** (*sodium sulphate*) and highly dangeorus, acidic *hydrogen chloride* gas. If the oil of vitriol is highly concentrated the gas produced can be passed through water to make a solution of the spirit of salt. The top diagram shows two ways in which one can prevent the water "sucking back" as the gas dissolves. If the *sulphuric*

acid is dilute, then some of the gas produced will form *aqueous hydrochloric acid* with the water in the vessel. If *hydrochloric acid* is the goal excess salt should be used, the resulting solution can be distilled to give *aqueous hydrochloric acid*, with perhaps some *hydrogen chloride* gas boiled off first, and mixed common salt and sal mirabilis left in the vessel. If excess *sulphuric acid* is used, the solution once distilled off will leave crystals of sal mirabilis.

AQUA REGIA, literally royal water, is one of the few reagents able to dissolve gold and platinum. It is a mixture of aqua fortis (*nitric acid*) and spirit of salt (*hydrochloric acid*). Best results are obtained using concentrated acids in the ratio 1:3 by volume. Aqua Regia looses its effectiveness so mix afresh for each use.

ALKALIS, from the Arabic *al-qali* (originally ashes of the saltwort plant), are salts of metals such as *sodium*, *potassium* and *calcium* that form bitter, caustic, slippery solutions with water. Alkalis are a subcategory of *bases*.

LIME or **QUICKLIME** is *calcium oxide* prepared by calcining limestone to a temperature of around 900°C (chalk is a soft, porous form of limestone). This lime-burning was practiced by the ancients for the production of lime mortar. **SLAKED LIME** is lime that has been slaked with water to form alkaline calcium hydroxide, this process generates great heat. A cooled suspension of fine slaked lime in water is known as **MILK OF LIME** and reacts violently with acids and attacks many metals. If slaked lime is heated above 580°C it decomposes to form lime and water. Limewash is pure slaked lime in water. When dry its calcite crystals produce a unique surface glow. Whitewash is made from slaked lime and chalk with other additives that may include glue, salt, ground rice or molasses.

POTASH, as its name implies, is prepared quite simply from wood ash. Mix the ashes with water, adding as much ash as the volume of water will reasonably allow. Stir every so often to avoid the sediment settling. Soluble potash will leach out from the other insoluble matter, then the solution can be filtered and evaporated to leave an impure potash. Calcining will purify it further, after which the salts can be dissolved, filtered and evaporated again. This can be repeated as many times as required.

SODA ASH like potash is prepared from the ashes of plants. Saltwort or seaweed such as kelp, which both have a high sodium content, can be used, but the result is likely to contain potash as well. A purer soda ash can be made by the Leblanc process which starts where our recipe for spirit of salt left off. Take sal mirabilis and melt and fuse it at red heat together with limestone and charcoal to make a black ash. Soda ash is leached from this once cooled, the other products of this reaction being insoluble. Soda ash can also occur naturally, particularly where seasonal lakes evaporate. One such example is the mineral **NATRON** which is a mixture of naturally occuring soda ash and sodium bicarbonate found on the edges of lakes in Lower Egypt. Natron was used in mummification.

LYE sometimes refers to solutions of soda ash or potash. It is also used for solutions of **CAUSTIC SODA** (*sodium hydroxide*) and **CAUSTIC POTASH** (*potassium hydroxide*), prepared by mixing milk of lime with soda ash or potash. This mixing produces a solution of caustic soda, or caustic potash, and a limestone precipitate. A medieval recipe for caustic soda crystals calls for 1 part each of slaked lime and soda ash with 7 parts water, boiled until the volume is halved, filtered and decanted 10 times then evaporated.

INDICATOR PAPER is used to test for acids and bases. A simple indicator paper can be made from red cabbage. Simmer a large finely chopped red cabbage until the water is a deep purple. Cool the water, add drops to an absorbent acid free paper then bake dry at a low temperature. The paper will turn pink for acids and blue or green for bases. Litmus paper, prepared from lichens such as *ochrolechia tartarea*, turns red in acids and back to blue in bases. 20th C. industrial recipes call for fermenting the lichen with potash, urine and lime. Simpler recipes recommend boiling ground lichen to extract the sensitive coloring material. Blue litmus paper is prepared by impregnating white paper with this litmus mixture. Red litmus is made the same way but a few drops of an acid are added to turn it red first.

SALTPETER, also known as **NITRE**, can be found naturally blooming on rocks in some parts of the world, hence *sal* (salt) *petrae* (of rocks). To obtain it from this source simply dissolve, filter and crystalize. It can also be made by mixing ruminant animal waste, rotting vegetable matter, mineral waste with high lime content, and potash, in a cone ventillated by layers of straw. This pile is "watered" weekly with the stale urine of ruminant animals. Once "ripe" the heap is left a while and the saltpeter will effloresce on its surface to be gathered, dissolved, filtered and evaporated to give raw saltpeter. If potash

is added during solution the *calcium* and *magnesium* impurities react to form more nitre and the precipitated carbonates which can be filtered out. The solution can be further clarified by adding a little glue, which forms a scum with impurities that can be skimmed off. Further purification can be made by dissolving an excess of this nitre in boiling water. As nitre is more soluble than the common impurities the solution will be almost entirely of nitre, the less soluble impurities remaining solid. This can be filtered or carefully decanted once settled, and evaporated to produce beautiful needle-like nitre crystals in floral arrays. **CHILI SALTPETER** (*sodium nitrate*) gets its name from a vast natural deposit in South America.

SAL AMMONIAC (*ammonium chloride*) forms on volcanic rocks near fume-releasing vents, and was made at the ancient temple of Jupiter-Ammon in Siwa on the Egyptian-Libyan border by burning camel dung and collecting the white residue that condensed from the smoke. Jabir's preparation of **SPIRITUS SALIS URINAE** is ammonium chloride made by heating a mixture of urine and common salt. Whenever organic matter containing *nitrogen* is submitted to destructive distillation, more or less **AMMONIA** is formed, notable sources being putrid urine, human hair, and the horns and hooves of oxen or deer – hence **SALT** or **SPIRIT OF HARTSHORN**. The literature shows some confusion as the white crystalline substance formed by the destructive distillation of these last two is often named Sal Ammoniac as well, although chemically it is ammonium carbonate. When sublimated Sal Ammoniac (*ammonium chloride*) forms ammonia and spirit of salt, both of which attack metals with a vengeance. Ammonium salts all give off ammonia when heated. Ammonia is highly soluble in water and forms *ammonium hydroxide*, a strong base with properties similar to alkaline solutions.

PHOSPHOROUS is a much more recent discovery but worth including here. It is first on record as being prepared in 1669 by Hennig Brandt, who heated the residue of evaporated urine (a normal person discharges just under one and a half grams of phosphorous per litre of urine) with powdered charcoal, condensing the vapors into a waxy mass that glowed in the dark. A more detailed description based on Leibniz's instructions is: Boil urine to reduce it to a thick syrup. Heat until a red oil distills up from it, and draw that off. Allow the remainder to cool and grind it finely. Mix the red oil back into the ground material. Heat that mixture strongly for sixteen hours. First white fumes come off, then an oil, then phosphorus. The phosphorus may be passed into cold water to solidify.

51

USEFUL RECIPES

CHARCOAL is produced by heating wood in the absence of the oxygen in air. It can also be made from bone. Charcoal burns hotter and cleaner than wood and is useful in smelting and forging for these reasons. Wood charcoal is mosty carbon and has been made since prehistory. Conical piles of wood were made with openings at the bottom and a central shaft for limited air flow, the pile was covered with turf or wet clay, and the firing started at the bottom of the shaft. Any suitably closed container of dry wood placed in a hot enough fire will produce charcoal with a little experimentation. It is important to allow just enough gap for gases to escape the container without allowing the free flow of air that would reduce the wood to ash. Twigs of vine or willow charcoal are popular for drawing. To produce small quantities follow the instructions for making bone black on page 46 replacing the bones with stripped twigs.

GUNPOWDER, or black powder, is a mixture of saltpeter, sulphur and carbon. Early Chinese recipes used equal weights to make a fast burning, but not explosive, powder. A composition that matches well the chemistry of the reaction is 15 parts saltpeter, 2 parts sulphur and 3 parts charcoal. Mix the ingedients while damp, using pressure to make a dense cake which can be broken into grains when dry. For most explosive results use refined saltpeter. Metal salts can add color to the explosion, e.g. sodium salts for yellow or orange, potassium salts for purple, and strontium salts for red. This is the basis of fireworks.

ANIMAL GLUE is made by simmering animal hides, tendons and hooves in water until they have broken down to give a thick glue which can be strained off. Take care not to heat too quickly or the mixture will burn and darken. The glue can be dried out to store. Mix with hot water 1:1 by volume to use. Glue made this way has been used for millennia and is good for carpentry – hide glue joints are repairable and reversible. Hide glue is kept liquid for working in a double boiler.

LEATHER is made by **TANNING** animal skins to keep them pliant even after they have become wet and dried out again. First scrape off all the fat and meat from the flesh side, then rub a strong solution of potash or lime into the fur side and leave for a couple of days until the fur becomes loose. Scrape the fur side with a sharp knife until clean. Traditionally the next stage is *bating*, an unassuming term for rubbing carnivore dung (usually dog) into the skin to break down its elasticity with enzyme reactions. Once the hide is no longer springy and lays flat the dung is washed out thoroughly. *Tannins* are leached from crushed oak bark in water, into which the skin is immersed for three days. After this the skin can be stretched out to dry and is ready. Skins can also be tanned with brains, each animal has just enough for its own hide. Clean the skin as above. Cook the brains in a small amount of water, squashing them with your hands to mix well. When the soup is as hot as you can still work with rub it into the flesh side then the fur side using your hands. Leave it for about seven hours then immerse the hide in water overnight. The water must then be worked out of the skin using a wooden wedge and a rounded stick. These tools help keep the skin stretched and loose while it dries. Smoking over a fire once dry helps prevent it stiffening again if it gets wet.

PARCHMENT, sometimes called **VELLUM**, is an animal skin treated with slaked lime and dried while stretched to produce a smooth surface for writing and painting upon. A summary of a 12th C. recipe is as follows: Stand goat skins in water for a day and a night, remove and wash thoroughly. Prepare a bath of milk of lime and immerse the skins, folding them on the flesh side, for a week (two in winter), agitating twice or thrice a day. Remove the skins and take off the hair. Make a fresh milk of lime bath, replace the skins and agitate daily for a week. Remove them and wash thoroughly. Soak in clean water for two days, then remove and tie the skin to a circular frame with cords. Dry, then shave the skin with a sharp knife and leave two days in the sun. Moisten and scour the flesh side with pummice powder. After two days repeat and fully smooth the flesh side with pummice powder while wet. Tighten the cords to flatten. Once dry the sheets are ready.

PAPER can be made with no chemical intervention, in which case the soaked, boiled, beaten and shredded plant fibers become a *mechanical pulp*. However, using an alkali breaks down the *lignin* from the *cellulose* fibers to give better results in a *chemical pulp*. Boil plant stems in an alkali such as slaked lime until the white fibers are left floating in a brown alkaline soup. Strain this pulp, soak in clean water, strain again and resoak. This pulp can be seived through a mold of wire mesh on a wooden frame to make sheets of paper.

52

IRON GALL INK is light-fast and burns into the page. Oak galls are swellings on oak trees caused by insect attack. Use the following by weight; 4 parts oakgalls, 1 part green vitriol (*iron sulphate*), 1 part gum arabic, 30 parts water. Grind the oakgalls finely and soak in half the water. Dissolve the green vitriol and gum arabic in the rest of the water, then mix both liquids. The instant black color will deepen if left for a month or two with occasional stirring. Excess iron salts will make an ink that turns brown at the edges while excess oakgall makes a weak black. **SEPIA** from Mediterranean cuttlefish and other such molluscs, is very long lasting with rich dark brown tones, but it is not light-fast. **INDIAN INK**, or sometimes **CHINESE INK**, is a *colloidol suspension* of carbon in water. Finely ground charcoal added to a thin gum arabic solution will make a simple Indian ink. The gum arabic binder also helps keep the carbon in suspension. Red Chinese ink replaces carbon with vermilion (see page 46) .

SOAP is made by *saponification*, the reaction between alkalis and animal or vegetable fats. Soaps made with caustic potash are liquid, while caustic soda makes soaps that are solid. The most popular fats used are lard, goat suet, beef tallow, olive oil and palm oil. A cold process with solid caustic soda or caustic potash allows an accurate approach at home. The following parts by weight can be used for 10 parts solid caustic soda, or 14 parts caustic potash, dissolved in 20 parts hot water; 72 parts beef tallow OR 73 parts lard OR 75 parts olive oil OR 71 parts palm oil. Gently melt the fat, if solid. The two liquids are best mixed at around 40°C, mixing too warm or at uneven tempertures are common errors. Pour the fat then the caustic soda into a suitable vessel and stir or shake vigorously. Mixed well the two liquids should not separate, if they do they must be reshaken or stirred. After a week the soap can be tested to see if it produces suds. Test for excess alkalinity with indicator paper.

GYPSUM is a common mineral. Calcined at around 150°C most of its chemically bound water is driven off to **PLASTER**. Mixing dry plaster powder with water reforms gypsum, firstly as a paste then expanding and hardening into a solid. A good preparation should be evenly mixed.

MORTAR was first used in ancient Egypt, as a mixture of gypsum and sand. Cement mortar is 1 part Portland cement, anything from 3 to 6 parts of sand depending on the strength needed (less sand is stronger), and water. Adding a coarse aggregate to this mix makes **CONCRETE**. Lime mortar is made by mixing 1 part quicklime and 2 parts fine sand with water. The quicklime, slaked in the mixture, hardens into limestone when exposed to air.

GLAZING PUTTY can be made by adding enough linseed oil to whiting (pure finely ground chalk) to make a paste. After it has dried it can be sanded. **LIME PUTTY** is slaked lime and water mixed thickly and left to stand to form a smooth paste.

The first known **ELECTRIC BATTERY** comes from Baghdad, dated to 250 BC, and consists of an earthenware shell, with an asphalt stopper pierced by an iron rod which is surrounded inside the jar by a cylinder of copper. In 1800 Alessandro Volta reinvented the battery by stacking layers of copper and zinc (the *electrodes*) separated by blotting paper soaked in brine (the *electrolyte*). A simple electric cell can also be made by placing an iron or galvanized (zinc coated) nail and a piece of copper wire (make sure they're not touching) as electrodes in a small jar of vinegar or lemon juice. The Daniell cell uses a zinc electrode in zinc sulphate solution above a copper electrode in copper sulphate, the two kept separate in a glass jar by the different densities of the solutions (see diagram). Another simple battery can be made with zinc and carbon rods immersed in sulphuric acid. Explosive hydrogen gas is produced so take care! Eventually the zinc rod will dissolve completely. Lead-acid batteries use electrodes of lead and lead oxide in sulphuric acid (diluted about 1:2 by weight with water). The lead and lead oxide react with the electrolyte to form lead sulphate and water while generating a current. This reaction can be reversed by passing a current through the battery, thus making the battery rechargeable.

DYES need to be fixed to a fabric's fibers so that they do not wash out. This often requires a **MORDANT**, the most popular being *alum* which mordants tea (rose), beetroot (gold), red onions (orange), madder (red), elderberries (lilac) and others. Mix a quarter of your fabric's weight of alum in enough water for the fabric. Wet the fabric in warm water then immerse in the mordant and heat for one hour stirring occasionally. Cool overnight. Boil your dyestuff in water for half an hour then add enough water to immerse the fabric. Heat for one hour or until the fabric's color is as you want it (it will lighten after rinsing and drying). Cool the fabric, rinse and dry. For a stronger color use more dyestuff, not more mordant. Indigo, or woad, require no mordant. Collect urine in a bottle or vat and stand it uncapped (or with a little exposure to air) in the sun until it has fermented. The strong smell of ammonia tells us it is ready. Add one teaspoon of very finely ground indigo per litre of urine. Stand in the sun for another day and you should have a pale green solution. Wash your fabric or wool with soap, rinse out the soap thoroughly, and place in the solution. Keep it submerged for 10 minutes then remove and squeeze out excess liquid. The fabric or yarn will turn blue in the air.

ZINC
ZINC SULPHATE
COPPER
COPPER SULPHATE
DANIELL CELL

INCENSE AND PERFUME

INCENSE at its simplest is the burning of woods, resins and herbs to make aromatic smoke. Its origins are lost in antiquity, although it probably first arose when woods such as sandalwood and agarwood were used on campfires. **Popular incense** ingredients that are quite readily available **include; Woods –** agarwood, cedar, juniper, pine, sandalwood, **and spruce. Resins** – peru balsam, copaiba balsam, benzoin, **camphor, copal,** dammar, dragon's blood, frankincense, **galbanum,** labdanum, mastic, myrrh, opoponax, sandar**ac, and** storax. Herbs – cassia, cinnamon bark, car**damom** seeds, cloves, coriander seeds, juniper, lem**ongrass,** patchouli, rosemary, common and white sa**ge, star** anise, sweet grass, thyme, vanilla and vetiver.

For non-combustible loose incense the ingr**edients** need simply be ground (if solid) and mixed. **It is best** to grind similar ingredients together then **mix them** once pulverized. If some of the ingredients **are liquids** the mix can be formed into pellets. Pur**e wood** charcoal once lit is ideal for burning loose **incense.** Self igniting charcoals contain saltpeter, the **inhalation of which is** strongly recommended against. Combustible **incense can be** made using makko, a naturally combustible **fine powder made** from the bark of an evergreen. Simply mix an**d knead with makko** and warm water then form into cones or roll onto thin sticks. Dry for a day and a night before lighting with a naked flame then fanning out the flame to leave a glowing tip. Some ingredients make combustible incense more readily than ot**hers, for example** sandalwood works well while frankincense is **difficult to** burn. Recipes and blends are best found **through** experimentation. First attempts at blending inc**ense** will be most successful with few ingredients as y**ou** learn which scents blend well with which other**s.** Some suggested blends are; cassia, clove an**d** sandalwood; cassia, frankincense, sandalwood and storax; juniper twigs, sweet grass and white sage; coriander seeds, frankincense, mastic **and** myrrh.

PERFUME applied to the body has as venerab**le a lineage as** incense. The ancient Egyptians made fragrant o**ils and waxes by** immersing petals and other fragrant materials i**n them. One can** also simply soak the source material in water to **make a perfumed** wash, roses to make rosewater for example. Materials for perfumery are obtained in many ways. Expression is used to squeeze out fragrant oils from citrus peel. Essential oils, (the

Spagyric Sulphur ⚕ of plants) are obtained by steam distillation. Enfleurage is a technique particularly useful for elusive scents, such as jasmine and tuberose which break down at the **temperatures of** distillation. Petals are placed on a bed of **scentless fat which** absorbs the fragrance, repeating this daily for **up to three months** the fat becomes saturated with scent. It is **then soaked in alc**ohol ☿ and warmed to extract the fragrance, **then c**ooled, filtered and the alcohol evaporated to **leave an ABSOLUTE. CONCRETES,** waxy solids **or thic**k liquids, are obtained using hydrocarbon **solvents** to extract the oils with other materials from **plants;** processed with alcohol they too yield **absolut**es.

Most raw fragrances are obtained from plants, **including** flowers, leaves, roots, seeds, fruits, woods, **barks,** resins and lichens. Fragrances from animal **sources** are also important. Musk from the Asian musk **deer, civet from** the civet cat, castoreum from the North **American beaver** and legendary ambergris, a waxy grayish **substance formed** in the intestines of sperm whales and found **floating at sea or wa**shed ashore, all have a long history of use and **are very good fixa**tives, adding a magical depth to blends and **extending their sta**ying power.

In recent times perfumery has developed the art of composing by notes. Top notes are the most volatile and immediately obvious, **middle or heart n**otes form the body of the fragrance, while base **notes are the** most long lasting and usually fix the other notes **for a more** enduring scent. The following essential oils and **absolutes** (abs.) are all readily available: Top Notes – **bergamot,** bitter orange, black pepper, blood orange, **cedar**wood, coriander, galbanum, lavender, lime, pink **grape**fruit, and rosewood. Mid Notes – clary sage, **gera**nium, jasmine abs., lavender abs., neroli, orange **flower** abs., rose abs., styrax, tuberose abs. and ylang **ylang.** Base Notes – ambrette seed, beeswax abs., **benzoin,** copaiba balsam, frankincense, labdanum, **oakmoss abs.,** patchouli, peru balsam, sandalwood, spruce **abs., tobacco ab**s., vanilla abs. and vetiver. A good starting point **for blending** natural perfumes is to use base, middle and top **notes in the ra**tio of 4:3:3. They should be blended in alcohol ☿ **then matured,** 95% grain or grape alcohol is ideal (vodka is not a suitable substitute for the amateur blender). 24 drops of base notes, 18 drops of middle notes, 18 drops of top notes, and 15 ml alcohol is a good guide to start from.

BHASMAS

In Indian alchemy there are methods for making metallic medicines called **BHASMAS** (Sanskrit – powder). The process makes compounds from metals by mixing them with plant ashes until no trace of the original metal remains. The aim is to connect intimately inorganic matter with organic so that its curative powers can be assimilated by the body, without toxicity. Zinc (which corresponds to ♃ like tin) is easy to work with and makes an excellent immune booster when made into a bhasma with turmeric.

Method: 1. Mix yogurt 1:2 with water in a bowl. 2. Melt a couple of grams of chemically pure zinc in a stainless steel spoon over a bunsen burner. 3. When just melted (do not allow to whiten) pour quickly into the yoghurt water. 4. Strain out the bits of metal and clean in water. 5. Repeat steps 2 to 4 six more times. The metal will become quite brittle. This completes the first stage of *Shodana* (purification). 5. Reheat the zinc in a large stainless steel spoon. 6. When partially melted add some ground turmeric. 7. Mix thoroughly (blow out if it catches fire). 8. As the burnt matter begins to whiten add more turmeric and keep mixing. 9. The metal should be amalgamated with the plant matter fairly thoroughly. If the contents of the spoon start to spill over before this stage is complete, set some aside and continue as before. All the zinc should be mixed until no visible particles remain. 10. Put the matter in an ovenproof porcelain pot and add a more than equal volume of fresh ground turmeric. 11. Add enough distilled water or rainwater to make a fairly loose paste. The inner Sulphur of the zinc should appear as a reddish color. 12. This is the first bhasma stage: Place the pot over the bunsen burner at full heat and cover with a lid to avoid oxidation. The lid should not be tight. Vapor must be able to escape, but not to enter. 13. As the mixture begins to smoke a red oil should start to appear on the surface. Calcine for at least 3 hours, by which time the bhasma will be dark gray. 14. Turn off heat and allow to cool a bit. 15. Make a fairly loose paste with as much fresh turmeric as there is mixture in the pot. 16. Add the paste to the hot bhasma mixture and stir in, adding more water to make a loose paste. 17. Repeat steps 12 to 16 for a total of 40 times. Fire is the great transformer and purifier, by the end of the work the bhasma should be a fine ash that fills the finger print when gently rubbed. Look at it under a microscope while shining a light on it to be sure that the original zinc is totally amalgamated. Any unamalgamated particles will reflect the light. If all is well, a tiny pinch of the bhasma mixed in water can be taken as a daily tonic.

FERMENTATION

FERMENTATION is the process by which plants produce alcohol. Chemically speaking, alcohol is produced from the interaction of yeast with plant sugars. The upper parts of most plants have sufficient amounts of airborne yeast to ferment naturally in the right conditions. The best spagyric essences are produced through natural fermentations. Method: 1. Finely chop a quantity of fresh herb. 2. Immerse the herb in up to five times its volume of good water in a non-metallic vessel. 3. Cover loosely and stand in safe place at a temperature of 16-28 °C. 4. Stir twice daily with a wooden spoon. Fermentation should start within three days. Once underway the plant matter rises to the surface and fizzes when stirred. This is *carbon dioxide* gas, a by-product of the process. 5. If fermentation fails to start, make sure you have some wine yeast activated with a little sugar on standby to add before the plant soup goes off (you will know by the bad smell). Alternatively, yeast and sugar can be added at the beginning to ensure successful fermentation. 6. Stir the brew twice a day and be sure to cover again (loosely) immediately. We want the heavy carbon dioxide to form a layer on the surface of the brew. This ensures the yeast produces plenty of alcohol and protects the brew from vinegar-forming bacteria, which can take over very quickly. If it has you can concentrate the vinegar by freezing the filtered brew. Since vinegar freezes at a lower temperature than water you can drain it from the ice and freeze it again, repeating till you have a strong vinegar – a very useful alchemical substance, especially in mineral work. Fermentation is complete once the brew stops fizzing and the plant matter sinks to the bottom. 7. Distill (gently) as soon as fermentation ceases. Distillation will separate both the Sulphur ♃ and the Mercury ☿ of the plant. 8. To keep the amount of water in the distillate to a minimum, while ensuring that all the ♃ and ☿ come over, taste the distillate from time to time. When it tastes insipid, cease fermentation. 9. If you have an alcohol meter, test the distillate to establish the alcohol content. It should be at least 16% ABV (alcohol by volume) if it is to keep well. 10. If the alcohol content is lower, pure ethanol (ca. 96% ABV or more) can be added. If the Sal Salis and Sal Sulphuris are extracted (see page 30) and added to the distillate you will have a spagyric essence. Such essences actually improve with age like fine wines.

PLANT PLANET CORRESPONDENCES

The planets govern all our functions on every level. Spagyric medicines can tone and harmonize our inner solar system. If our inner Venus needs a boost, for example, a spagyric tincture of yarrow or another Venus herb may help. Here are some of the principle qualities and rulerships of the planets and lists of some of the plants they rule.

☉ Sol is vitality, consciousness, the individual soul. It corresponds to Sulphur, the hot, dry, masculine principle, the active, engendering seed, called the Father of the Stone. The Sun's influence is benign, but in excess can engender pride and egoism. Without the cooling, moistening influence of the Moon, it can be arid and burning. The Sun rules the mind, energy, willpower; physiologically the heart, eyes, circulation and health in general.

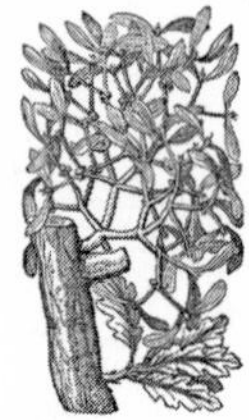

Angelica archangelica: angelica - *Anthemis nobilis*: Roman chamomile - *Boswelia sacra*: frankincense - *Bursera fagaroides*: copal - *Calamus aromaticus*: calamus - *Calendula officinalis*: marigold - *Cinnamomum ceylanicum*: cinnamon - *Citrus specia*: all citrus trees - *Droserae*: all sundews - *Echium vulgare*: viper's bugloss - *Euphrasia officinalis*: eyebright - *Fraxinus excelsior*: ash (with ♃) - *Helianthus anuus*: sunflower - *Hydrastis canadensis*: golden seal - Hypericum perforatum: St. John's wort - *Juglans regia*: walnut tree - *Matricaria chamomilla*: German chamomile - *Myrrba*: myrrh (with ♃) - *Oryza sativa*: rice - *Paeonia officinalis*: peony - *Rosmarinus officinalis*: rosemary - *Ruta graveolans*: common rue - *Sorbus aucuparia*: rowan tree - *Syzygium aromaticum*: clove tree - *Viscum album*: mistletoe (with ♃ and ☽) - *Vitis vinifera*: grapevine (with ♃ and ☽) - *Zingiber officinale*: ginger.

☽ Luna rules the emotions, instincts, the subconscious. It is feminine, motherly, nurturing, reflective, changeable. It influences fertility, growth, and conception. The waters of the oceans, the sap in plants, and all bodily fluids are influenced by the Moon, as witnessed by the tides and the menstrual cycle of woman. All things grow in rhythm with the Moon. It rules dreams, emotions, sensuality, intuition. Its dark side is the unconscious, the wilder, baser instincts. It is the bride of the Sun, Diana, the lunar goddess of the Greeks. Luna is the White Queen, the White Lion, the Elixir of Immortality, which converts metals into silver. It is also Mercury to the Sun's Sulphur; the cold, moist, passive, feminine principle

Mistletoe

Cabbage

that receives the seed of Sulphur and bears the hermaphroditic child. Physiologically it rules the stomach, cerebellum, female reproductive organs, lymphatic system and pancreas.

Acanthus mollis: acanthus - *Agnus castus*: monk's pepper - *Bellis perennis*: daisy - *Brassicae*: the cabbage family - *Cardamine pratense*: cuckoo flower - Cucumis sativus: cucumber - Cucurbita pepo: pumpkin - *Curcuma longa*: turmeric - *Galium aparine*: cleavers - *Iridaceae*: irises - *Lactuca sativa*: lettuce - *Myristica fragrans*: nutmeg - *Nymphaea alba*: water lily - *Ruta lunaria*: moonwort - *Salices*: willows - *Saxifraga*: saxifrages - *Sedum telephium*: orpine - *Stellaria media*: chickweed - *Tilia*: linden (lime) trees - *Veronica officinalis*: speedwell - *Vinca minor/major*: periwinkle - Most water plants.

☿ Mercury is the fastest moving planet; the quicksilver messenger service mediating between Above and Below. It rules mental processes, travel, communications, language, writing, adaptability and the intellect. It shares the same ambivalent qualities as Hermes/Thoth/Mercury. As a planetary entity Mercury has a puckish, trickster tendency that exposes falsehood and conceit. Being androgenous, and containing all opposites, he is therefore a free operator, independent of a polar opposite, although he has an antagonistic relationship with Saturn. The planet Mercury is not to be confused with the Mercury of the Sulphur-Mercury duad.

Physiologically Mercury rules the nervous system, hearing, tongue, throat, lungs, coordination, the spinal cord.

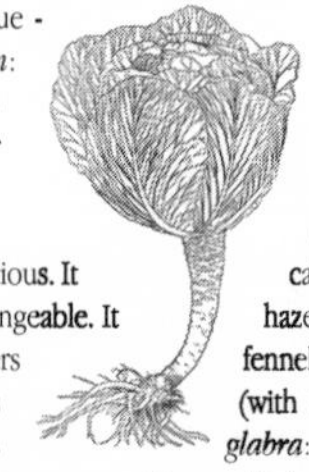

Acacia: acacia species - *Anethum graveolens*: dill - *Artemisia abrotanum*: southernwood - *Atropa mandragora*: mandragora (with ♄ and ☽) - *Bryonia alba*: bryony - *Calamintha montana/arvensis*: calamint - *Carum carvi*: caraway - *Corylus avella*: hazel - *Daucus carota*: carrot - *Foeniculum vulgare*: fennel (with ♃) - *Geranium robertianum*: herb Robert (with ♀ and ♂) - *Ginkgo biloba*: ginkgo - *Glycyrrhiza glabra*: licorice - *Lavandula vera*: lavender (with ♃ and ☉) - *Majorana hortensis*: marjoram - *Marrubium vulgare*: white horehound - *Mercurialis annua/perennis*: annual/perennial mercury - *Morus*: mulberry species - *Mytrus communis*: myrtle - *Origanum vulgare*: oregano - *Petroselinum hortense*: parsley - *Pimpinella anisum*: anise - *Satureja hortensis*: savory - *Scutellaria lateriflora*: skullcap - *Solanum dulcamara*: bittersweet (with ♄) - *Valeriana officinalis*: valerian.

♀ Venus is the planet of affection, art, and music. Venus helps mediate between opposites and to integrate diverse elements into harmonious balance. Venus/Aphrodite is the Goddess of Love, but to the Egyptians, Indians and Hebrews is masculine. To the Indians he is Sukra, who, like Thoth, is a teacher and physician. Legend relates that Sukra possessed the Elixir of Immortality. The influence of Venus is entirely benign, but if badly aspected can promote sexual and sensual excess. Venus rules the complexion, breasts, thymus gland, fertility, kidneys, inner sexual organs, blood and cell formation and the sense of smell.

Achillea millefolium: yarrow - *Ajuga reptans*: bugle - *Alchemilla vulgaris*: lady's mantle - *Aquilaria agallocha*: agarwood, oud (with ♃) - Aquilegia vulgaris: columbine - *Arctium lappa*: burdock - *Artemisia vulgaris*: mugwort - *Betula*: birch trees - *Castanea sativa*: sweet chesnut (with ♃) - *Leonurus cardiaca*: motherwort - *Menthae*: all kinds of mint - *Nepeta cateria*: catnip - *Persica vulgaris*: peach - *Primula officinalis*: primrose - *Pyrus communis*: pear - *Pyrus malus*: apple - *Rosa*: roses (with ♃) - *Santalum*: sandalwood (with ♀) - *Saponaria officinalis*: soapwort - *Solidago virga aurea*: goldenrod - *Tanacetum parthenemium*: feverfew - *Thymus vulgaris*: thyme - *Triticum sativa*: wheat - *Verbena officinalis*: vervain - *Vetiveria zizaniodes*: vetiver - *Viola odorata*: violet.

Tobacco

♂ The red planet represents the intensely masculine, active, dynamic principle. Its effects are intensifying, accelerating and violent. As the god of War, Mars is traditionally seen, along with Saturn, as a malign body at conflict with the other planets. The negative aspects of Mars include ruthlessness, destruction and brutality. Positive aspects include determination, willpower, courage and passion. Mars rules the muscular system, sex organs, and blood formation. Paracelsus tells us that Mars governs the polarity between the brain pole and the sexual organs, the conduit for kundalini, in Indian alchemy.

Allium cepa: onion - *Allium sativum*: garlic - *Berberis perennis*: barberry - *Bryonia dioica*: white bryony - *Capsicum*: hot peppers - *Carduus benedictus*: blessed thistle - *Cochlearia armoracia*: horseradish - *Coriandrum sativum*: coriander/cilantro (with ♀) - *Crataegus*: hawthorn - *Gratiola officinalis*: hedge hyssop - *Humulus lupus*: hop - *Mentha pulegium*: pennyroyal - *Nicotiana tabacum*: tobacco - *Pausinystalia yohimbe*: yohimbe - *Pinus*: pines - *Plantago lanceolata*: ribwort plantain - *Plantago major*: plantain - *Potentilla tormentilla*: tormentil - *Rheum palmatum*: Chinese rhubarb (with ♃) - *Rubia tinctorum*: madder - *Smilax utilis*: sarsaparilla - *Turnera aphrodisiaca*: damiana - *Urtica dioica/urens*: stinging nettles.

Comfrey

♃ Jupiter is by far the largest of the visible planets. In contrast to the restrictive and inhibitory qualities of Saturn, Jupiter is expansive, generous, warm and jovial. He is the fire of Nature, the warmth in all things. In mythology he is the lusty, self-indulgent king of the gods. This tendency to excess is Jupiter's main weakness. As with all the planets, qualities can be inverted if badly aspected by other planets. Jupiter presides over law, harmony and religion. Physioligically Jupiter governs the liver, the immune system, circulation, digestion, thighs, feet and teeth.

Acer: maple, sycamore - *Aesculus hippocastanum*: horse chestnut - *Agrimonia eupatoria*: agrimony - *Agropyron cannium*: couch grass - *Amygdalum*: almond - *Anthriscus cerefolium*: chervil - *Arnica montana*: arnica - *Betonica officinalis*: betony - *Castanea sativa*: sweet chesnut (with ♀) - *Foeniculum vulgare*: fennel (with ♃) - *Fraxinus ornus*: manna, flowering ash - *Fumaria officinalis*: fumitory - *Gentiana lutea*: yellow gentian - *Hepaticae*: liverworts - *Jasminum*: jasmine - *Melilotus officinalis*: melilot - *Melissa officinalis*: lemon balm - *Ocimum basilicum*: basil (with ♄) - *Panax ginseng*: ginseng - *Prunus armeniaca*: apricot tree - *Salvia officinalis*: sage - *Solanum lycopersicum*: tomato - *Tanacetum vulgare*: tansy - *Taraxacum officinale*: dandelion - *Rheum palmatum*: Chinese rhubarb (with ♄) - *Verbascum thapsiforme*: mullein.

♄ Saturn is the Guardian of The Threshold between the material and the spiritual world, where the descent into matter begins. The most distant and slowest moving of the visible planets, it represents restriction and inhibition. As the Lord of Order, self-knowledge, and discipline, Saturn can be a severe taskmaster. It governs all crystallizing and hardening mechanisms, ruling the skeleton and ageing processes. Saturn is presented as the skeletal figure of Death or Old Father Time with his scythe, mercilessly cutting down the old, useless or unworthy. This applies to the dross or impurities of the alchemist's materia, including the corresponding aspects within himself. Saturnine conditions include rheumatism, depression and chronic diseases.

Atropa belladonna: belladonnna (with ♄) - *Cannabis sativa*: hemp - *Cinnamomum camphora*: camphor - *Cupressus sempervirens*: cypress - *Dryopteris filix-mas*: male fern - *Epilobium angustifolium*: willow herb - *Eriodictyon californicum*: yerba santa - *Equisetum arvense*: horsetail fern - *Fagus silvatica*: beech - *Foenum graecum*: fenugreek - *Fumaria officinalis*: fumitory - *Hedera helix*: ivy - *Hordeum*: barley - *Hyoscyamus niger*: henbane - *Ilex aquifolium*: holly - *Papaver*: poppies (with ☽) - *Piper Methysticum*: kava kava - *Polygonatum officinale*: Solomon's seal - *Symphytum officinale*: comfrey (with ♃) - *Taxus baccata*: yew.

Astrological Hours

When making spagyric preparations it always helps to reinforce the planetary signature by performing or beginning each process or stage on the correct planetary day and, where possible, during the appropriate planetary hour. Traditionally the day is divided into a number of planetary hours, which vary according to different sytems. The planetary sequence of these hours always corresponds to the heptagram on page 22 when followed anti-clockwise.

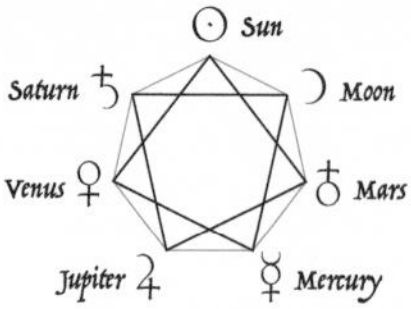

Thus the hour of the Sun ☉ is followed by the hour of Venus ♀, then Mercury ☿, the Moon ☽, Saturn ♄, Jupiter ♃ and Mars ♂. The sequence then repeats. The first hour of each day corresponds to the day in question: the first hour of Sunday is a Sun hour. Views differ as to when the day is deemed to begin and on the length of planetary hours. According to Celtic, Kabbalistic and Islamic traditions the day begins after sunset. Sunday therefore starts on "Saturday" evening. Systems can be fixed or flexible. In the fixed Kabbalistic system the day always starts at 6.00 p.m. In a flexible system the day begins after sunset, whatever the hour. Other traditions hold that the day starts at sunrise. Again, fixed and flexible systems can apply. In these systems each hour lasts 60 minutes. Alternatively the periods of daylight and darkness can be divided into 12 "hours" which vary in length according to the time of year – short daylight "hours" in winter, long ones in summer. A popular system amongst Western alchemists uses seven equal periods, starting at midnight (*below*), the second period matching the day. The advantages of this system are that it is constant, while sunrise tends to occur during the hour that rules the day.

Hours	Sunday	Monday	Tuesday	Wednesday	Thursday	Friday	Saturday
0:00 to 3:26	♂	☿	♃	♀	♄	☉	☽
3:26 to 6:52	☉	☽	♂	☿	♃	♀	♄
6:52 to 10:18	♀	♄	☉	☽	♂	☿	♃
10:18 to 13:44	☿	♃	♀	♄	☉	☽	♂
13:44 to 17:10	☽	♂	☿	♃	♀	♄	☉
17:10 to 20:36	♄	☉	☽	♂	☿	♃	♀
20:36 to 0:00	♃	♀	♄	☉	☽	♂	☿

Alchemical Symbols

THREE PRINCIPLES

🜍	🜔	☿
Sulphur	*Salt*	*Mercury*

THE FOUR ELEMENTS

🜂	🜁	🜄	🜃	+
Fire	*Air*	*Water*	*Earth*	*The Elements*

PLANETS & METALS

☽	☿	♀	☉	♂	♃	♄
Moon	*Mercury*	*Venus*	*Sun*	*Mars*	*Jupiter*	*Saturn*
Silver	*Quicksilver*	*Copper*	*Gold*	*Iron*	*Tin*	*Lead*